SAFECRACKER

SAFECRACKER

A Chronicle of the Coolest Job in the World

DAVE McOMIE

LYONS
PRESS

Guilford, Connecticut

An imprint of The Rowman & Littlefield Publishing Group, Inc.
4501 Forbes Blvd., Ste. 200
Lanham, MD 20706
www.rowman.com

Distributed by NATIONAL BOOK NETWORK

British Library Cataloguing in Publication Information available

Library of Congress Cataloging-in-Publication Data available

ISBN 978-1-4930-5851-8 (cloth : alk. paper)
ISBN 978-1-4930-5852-5 (electronic)

♾™ The paper used in this publication meets the minimum requirements of American National
Standard for Information Sciences — Permanence of Paper for Printed Library Materials,
ANSI/NISO Z39.48–1992.

Contents

CONTENTS

Foreword

WHAT COMES TO MIND WHEN YOU HEAR THE WORD "SAFECRACKER"? Maybe a thug with more tattoos than teeth, pounding on a safe with a sledgehammer, and cussing through a dangling cigarette. Or maybe an action hero like Tom Cruise, contorting his body through a maze of infrared laser beams while rappelling over a towering safe full of flawless diamonds. The last thing you'd expect is a compact twenty-something with a Beatles haircut, a suitcase full of tools, and a philosophy book. But in early 1986 that is exactly who I discovered.

Back then, I was the editor and publisher of *The National Locksmith*, a trade magazine for the security industry. I was always on the lookout for locksmiths and safe technicians who could translate their technical know-how into readable articles. One day, while going through the mail, I discovered a handwritten letter from one Dave McOmie, offering to write a recurring column on safecracking. He was so confident and so clearly knowledgeable on the topic that I sent him a plane ticket. I just had to meet this guy.

With no idea what Dave looked like, I found myself standing at O'Hare Airport, holding a sign that said simply "MCOMIE." I scanned the bustling crowd at Arrivals until a lone traveler caught my gaze, smiled and waved, and strode over to shake my hand. I was incredulous that a seasoned safecracker could be so young, but we hit it off immediately, and the rest is history.

Dave took the industry by storm, and his vivid descriptions of life as a master safecracker made him our most popular columnist, virtually overnight. Everyone enjoys the adventures of a guy with the coolest job in the world!

His occupation is ruled, quite unforgivingly at times, by precise measurements for exploiting design vulnerabilities in safes and vaults. You either have these measurements — known as *drill points* — or you don't. Dave has them, to the enduring delight of his reading audience.

He hosted a monthly column in *The National Locksmith* for more than three decades before devoting himself exclusively to *The International Safecracker*, his quarterly journal for working professionals. He has authored two dozen technical books, taught thousands of students at his legendary Penetration Parties, and traveled far and wide to open the most difficult safes and vaults in the country.

In the small community of locksmiths and safecrackers, Dave is a rock star, albeit a reluctant one. Averse to both crowds and chitchat, he's a bit of an introvert who has never embraced his celebrity status. Indeed, he would rather read than mingle, as I have witnessed firsthand.

We were at the annual Associated Locksmiths of America convention, announcing the release of one of Dave's early books. His hotel reservation fell through, and there were no rooms available. It was late evening, so I offered him the second bed in my suite. When it was time to turn out the lights, he slapped on headphones and put a tape in his Walkman. He was listening to the audio track of a debate over the morality of capital punishment.

The following morning, a little before dawn, he was sitting up in bed, book light attached to *The View from Nowhere*, a dense and difficult book by philosopher Thomas Nagel. This pattern repeated all three days. Dave did his duty on the convention floor, signing books and posing for photographs. But on his own time, he was either immersed in a book or discussing a serious issue with one or two people, well away from the limelight.

This unusual combination of talents and proclivities made Dave a rare breed from the get-go. It also suited him, quite uniquely, to produce this autobiographical account of life as a professional safecracker — the first and only memoir of its kind.

You're about to meet the slightly reclusive, deeply reflective safe-cracker I've known for more than thirty years. I never thought this private man would go public, but I'm thrilled he did. You will be too.

Buckle up!

Marc Goldberg
Former Editor/Publisher
The National Locksmith

Author's Note

All people and events in this book are real. Where feasible, I have divulged the names of persons, banks, and other companies that engaged my services. In a few places, however, I have honored requests for anonymity or privacy. And for the sake of narrative flow, I have compressed certain time sequences and recreated dialogue that comports with my best recollection of those conversations.

MONDAY

Chapter One

The Call from Vegas

Sparks fly off the drill bit I'm pressing gently against the grinding wheel. Dull bits are dead bits that are either resurrected or unceremoniously buried, and this one is being brought back to life. The smell of burnt metal tells me to ease up. I'm checking the tip for sharpness and symmetry when my vest pocket vibrates twice — a text. Through smudged safety glasses, I read a single question on the screen: *Available to drill a bank vault?*

It's Mike Hitchcock from Vegas. Hitch is a field technician with Diebold, my biggest and best client. They maintain cash safes, vaults, ATMs, and safe deposit boxes in banks and credit unions in all fifty states.[1] I turn off the grinder, text him "yes," and ask if he prefers tomorrow morning or afternoon.

Hitch's answer catches me by surprise: *Must be today, Dave. Call me when you break free.*

Whoa. I rarely open a bank vault same day, especially when it requires a plane ride. A day's notice is standard. This provides a little time to review the job and plot an effective attack, including Plan B in case things go south.

Hitch gives me the lowdown over the phone. He's serious — this vault needs to be open before midnight tonight. It's not at a brick-and-mortar branch but at a major banking chain's private currency center on the outskirts of town. Bank personnel dialed the correct combination this morning, but the vault's handle wouldn't turn. They soon discovered why: the timelock had been overwound by approximately twenty-four hours.

A timelock is an ingenious device that prevents a bank vault from being dialed open after hours. Inside the timelock are three small clocks, each with a square post that fits perfectly in a winding key. Two bank employees approach the vault at closing time. One person winds time on all three clocks, then the other double-checks the number of hours wound.

Once they close and lock the vault door, it cannot be opened until time winds down on at least one clock. (Three clocks provide double redundancy, in case one malfunctions.) This ultra-secure method of locking — using combination locks *and* a separate timelock — has deterred potential kidnappers and rogue employees for the past century and a half. Bank personnel can't open the vault after hours for any reason, not even at gunpoint. And there's no such thing as a timelock override other than in the make-believe world of movies and mystery novels.

A Diebold three-movement timelock. Once time is wound on all three movements and the vault is locked, it cannot be opened until time runs down on at least one movement/clock.
Source: Author photo.

Overwinds are the result of a mistake going undetected. The person double-checking the winder can become progressively less diligent about bending over and peering through the magnifier to confirm the number of hours on each clock. The winder almost always gets it right, so diminishing attention over time is understandable.

But then one day, perhaps after *years* without a mistake, the tedium takes its toll and too many hours are wound on all three clocks. The person double-checking, who by this time has become more of a rubber stamper, fails to notice, and the vault is closed and locked. The next morning, it won't open. This happens somewhere in the United States every week, most often on Monday.

I don't know how this particular overwind occurred, and I'm not interested in embarrassing an overworked bank employee. What I do know is that the bank will face severe financial penalties if their clients' money isn't in transport by midnight. Hence the hurry.

Hitch asks the obvious question. "Do you want to chance it, with the deadline so soon?" I hesitate. If I fly down and get it open today, I'll send a healthy invoice to Diebold for services rendered. But if the clock ticks one second into Tuesday and the vault isn't open, I'll have to put my tools away and go home with my tail between my legs. From the very beginning of our relationship, which goes back more decades than I care to count, my pitch to Diebold has always been that I charge for success, not failure. And I have never *not* charged them.

I'm wary of putting my perfect record at risk. A flight delay, unexpected tool breakage, unknown issues with the vault — any of these things could push me past the deadline. But I'm drawn to the challenge. Can I fly to Vegas and get the job done before midnight? Like a bee in nectar's thrall, my inner naysayer is muzzled, unable to decline. "What the heck, Hitch, let's go for it. I'll book a flight and text you my itinerary."

"Sounds good, Dave. See you soon."

Half a nanosecond after we hang up I realize that I didn't ask about make and model — crucial to know so I can pack the right equipment. I text Hitch. He doesn't know, but a man on the inside will get us that information, hopefully with a cell phone photo of the vault door.

At midnight on January 26, 1876, John Whittelsey was tortured in his home until he divulged the combination to the vault at Northampton National Bank in Massachusetts, where he worked as head cashier. The robbers proceeded to the bank, where they dialed opened the vault and made off with eight hundred grand. Fearing an epidemic of similar crimes, bankers quickly turned to timelocks, which solved the problem. Having the combination was no longer enough; time had to wind down before the vault could be opened. To deter potential robbers, banks began to post signs, alerting the public to the fact that the vault cannot be opened after-hours due to timelock. Such signs are still in use today. Whittelsey's ordeal was recounted later that same year, in a short, thirty-page book sporting the longest, most descriptive title and subtitle I have ever seen: *The Greatest Burglary on Record: Robbery of the Northampton National Bank: the Cashier Overpowered at Midnight, Tortured, and Forced to Give the Combinations to the Vault and Safe: Eight Hundred Thousand Dollars in Money and Bonds Stolen: $25,000 Reward.* The author is anonymous.
Source: Courtesy of the General Society of Mechanics and Tradesmen.

I hustle into my home office, wake up the iMac, and book a flight. I need to leave in about an hour, so I gag down an energy drink and head to the garage, still puckering. At three bucks a teeny-tiny bottle, the darn stuff should taste better.

My tools are still in a messy pile on the workbench. When this call came in, I was doing a little tool maintenance — sharpening or replacing dull drill bits, smoothing the ragged tips of my pin punches, straightening my wire probes true again, recharging the batteries in my electronic devices.

It's time to repack everything into the wheeled carry-on suitcase I work out of. This case has to be checked, due to its contents and weight, so it must be well built and durable. Over the years, I tried a half dozen different cases that came back with broken wheels or handles after just a few flights. Then I found this BMW-labeled carry-on, which is lightweight but built like a Panzer. It's almost TSA-proof.[2]

The Beemer case is half-filled when Hitch texts me the answer. His contact is no longer onsite and can't get a photograph, but he did identify the bank vault as a Diebold Rotary Wedgelock — a thick door requiring long drill bits. Good to know. I finish packing and make sure there are plenty of twelve- and eighteen-inch bits, and equally long wire probes as well.

I close the case and put it on my scale. We're four pounds overweight, so I remove a bank bag full of short carbide bits and another one half-filled with diamond bits. Diamond bits won't be needed, but I grab a couple of short carbides to use during the first few inches of penetration. Now at a svelte forty-nine pounds, we're in good shape with a little leeway for the airport scale.

I enter full-on departure protocol: wheel the Beemer to the garage door and park it next to the case containing my portable drill press. Dial open my tallest safe and pull out the aluminum briefcase that houses a dozen of my medical scopes.

Behind the door to my office is another carry-on suitcase preloaded with toiletries and a change of clothes. I lay the briefcase on top of the clothes, zip the carry-on closed, and slide my laptop case over the handle. Ready to rumble.

Almost. There are a few minutes to spare, and I'm hungry. In the fridge is my wife's leftover spaghetti from last night — much better than airplane food. I call to let Valerie know my itinerary, she wishes me safe travels, and I'm off.

Chapter Two

Your Host

As a kid, I was enamored with Alexander Mundy, the debonair safecracker from the 1960s television series, *It Takes a Thief*. So enamored, in fact, that I enrolled in a locksmithing home-study course and later landed an apprenticeship at a local shop. I found my passion, then my job. It turned into a career.

My interest in the key-cutting side of the business faded as I became familiar with safes and vaults. The supreme challenge was opening them, and this was reflected in the price difference: we charged five bucks to rekey a door lock, and five *hundred* to drill a tough safe. It was like learning the greatest magic trick of them all, and getting paid for it.

Bitten badly by the safe and vault bug, I left general locksmithing early on and went to work as a professional safecracker, specializing in difficult lockouts, and have spent much of my working life crisscrossing the country, opening bank vaults and high-security safes.

People tend to have one of two reactions upon learning of my occupation. Faces often light up as they envision *The Italian Job* or *The Score*. Don't I wish! Safecracking in the movies is usually more fun to watch than it is in the real world.

Others look at me a bit cautiously, unsure if what I do is legal. I try to be crystal clear here. I don't open safes in the wee hours under the glare of a flashlight, but during work hours, at the behest of bankers and jewelers anxious to regain access to their stock-in-trade.

Genuinely befuddled, people will often follow up by asking why anyone would ever *need* to have a safe cracked. As it turns out, there

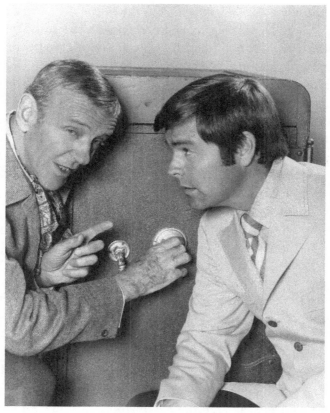

Like father, like son: Fred Astaire and Robert Wagner as Alistair and Alexander Mundy in *It Takes a Thief*. The show is fun, escapist entertainment, but its central premise — that the best safecrackers are suave, debonair thieves — is false on all three counts.
Credit: Photofest.

aren't a thousand reasons why people retain my services. There are exactly four:

One, the combination is unknown. A locked safe was purchased at an auction or on Craigslist. Or the numbers were forgotten or went to the grave with the only person who knew them. These can be fun, especially when a client hits the jackpot.

Two, the combination is known, but the safe won't open. Something has gone wrong. Depending on the problem, these can be quite challenging. The fifty most nightmarish lockouts I've ever dealt with were all bizarre, undiagnosable malfunctions.

Three, bungling burglars have made a mess. The safe cannot be opened normally because the dial, handle, and/or hinges have been hammered, cut, or burned off in a botched attempt to misappropriate money and jewels.

Four, operator error. The user forgot how to dial the combination, or failed to seat the key properly, or didn't turn the handle far enough. Or they overwound the timelock — today's scenario in Las Vegas.

Each day brings a fresh challenge. Somebody is locked out of their safe or vault, and my job is to get them in. It's an unusual way to put food on the table, I know, but it's challenging in ways that few things are and deeply rewarding. I can't imagine doing anything else.

The only other paying job I ever had was playing lead guitar for singer Leon Smith in the early 1980s. I loved guitar (still do), and being in the band was a blast, but I blew my ears out in a few years and had to bail before the ultra-loud stage amplifiers rendered me deaf.

At another point I wanted to teach philosophy, but in graduate school it became apparent that the demand for academic philosophers was waning. This trend eventually reversed, but by then I had left the program with a master's degree and returned to my roots, this time for good.

Although I didn't pursue a professorship, studying philosophy changed my life. It made me a more thoughtful person by forcing me to sit still and think, and by requiring an honest summary of the opposing side in an argument. Do this enough times, and you'll gain an appreciation for positions you previously thought were obviously wrong.

Regarding life's big questions, my interest is as strong as ever, but conclusions remain elusive. I haven't a clue how organic life arose, or if natural selection working on accidental variations in the genome really can account for the panoply of lifeforms on our planet. I don't know if we're mere molecules in motion or something more, or how free our will actually is. And while I won't pretend to *know* whether or not God exists, I wish I did.

I'm agnostic, yes, but with an abiding appreciation for life's many gifts. I like cheap wine from Trader Joe's and expensive steaks from Japan. I love playing ping-pong and chess with my younger kids and Words With Friends with my wife and older kids.

If I go more than a few days without reading a serious book or cracking a serious safe, I get anxious and agitated. So I read and crack. A lot.

And when I can't do either, I listen.

Chapter Three

From Airport to Airport

HEADING WEST TOWARD THE AIRPORT, I TURN ON THE RADIO. BUT before I can select a station, my phone rings through the speaker system. It's Shane Ellison, calling from North Carolina. With his natural smile and relaxed southern manner, Shane is a pleasure — the younger brother I never had. He's also an extraordinarily talented safecracker. Of all the students in the countless classes I've taught over the past three decades, he's one of a small handful to have been invited into the inner circle, where defeat methods for the most secure containers in the world are shared and discussed.

Shane had an intriguing job earlier today, assessing the damage to a vault door that was attacked during the burglary of a large jewelry store just before dawn. The crooks got away with millions.

"Wow," I reply. "How did they gain entry into the vault room? Did they core a hole through the wall or ceiling?"

"Well, that's the strange part. The police detective and insurance investigator are both convinced it was a professional job on the door itself."

"What?" I reply. "Was the door opened?"

"Yes sir."

Shane knows as well as I do that there have been very few successful burglaries on bank vaults in the United States. And nearly all of them involved coring a man-size hole in the vault wall or ceiling, rather than opening the door itself.[1]

"Did you examine it?"

"Yes sir."

"Did the burglars drill a hole through the door into the combination locks or the timelock?"

"No. The only damage to the door, Dave, was a burn on the outer skin from an oxyacetylene torch."

"You mean the penetration didn't go all the way through? They burned through the thin piece of stainless steel and stopped?"

"That's right."

"Shane, this sounds like an inside job. The perps probably had the combination, opened the vault, and then tried to cover their tracks to make it *look* like a burglary. And from your description, they didn't do a very good job of it."

"That's what I told the detective and the investigator. I showed them that penetrating one-eighth of an inch of a sixteen-inch-thick vault door couldn't have opened it. But I didn't arrive until they were finishing up, and they had pretty well made up their minds by then. What's even weirder is that they missed another glaring piece of evidence."

"There were two carts sitting side by side in the vault, one filled with new jewelry, the other with repaired jewelry. The burglars took *only* the boxes containing new jewelry. They didn't touch anything on the repair cart. Obviously, they knew which was which."

"Nice catch, Shane. Somebody wanted to avoid the hassle of dealing with a bunch of upset clients."

"Exactly. I can't believe this is going to go down, officially, as a successful burglary. They're going to get away with it."

"You're probably right. Hey, what model of vault door is it?"

"A Rotary Wedgelock."

"You're kidding."

"Serious as a heart attack."

Traffic slows as I approach the airport exit.

"I'm about to get on a plane for Vegas, Shane. Guess what model vault I'm drilling tonight?"

"A Rotary Wedgelock? No way!"

"Yes way. I'd better go, mister, so I don't miss my turn. Say hi to Ann and the boys for me."

"Will do, Dave. Good luck tonight."

A few years ago, Shane's oldest son became the youngest student ever to open a burglary-rated safe at one of my classes. Jacob had recently celebrated his ninth birthday. I don't anticipate that record being broken anytime soon.

At the short-term garage, I spiral up the ramp and find a parking spot. From the south side of Level Four at PDX it's a direct walk across the southern skybridge to an escalator that drops you right next to the check-in counter at Alaska Airlines.

Standing in the fastest of the slow lines through security, I feel slightly agitated. Did I forget something? An oversight would be inexcusable and beyond embarrassing. It's too late to run home; even so, I take a quick mental inventory of the tools that were just checked at the counter. The process is mildly therapeutic, and it diverts my attention from the monotonous tension that permeates modern airports.

As expected, nothing is missing. Maybe it's nerves — few jobs rattle me, but today's time constraint is a little out of the ordinary.

I walk through the body scanner and return to the conveyor. My laptop is there but my aluminum briefcase is not. *Crap.* I look back at a poker-faced TSA agent sitting on a padded stool, leaning forward and staring intently into his monitor. He glances at me over his wire rims and turns to a colleague who listens to his instructions.

Suspect luggage in tow, she confirms ownership and motions for me to follow. I grab my laptop and walk over to the stainless steel table upon which my briefcase now rests. The problem isn't the case, but its contents: a dozen medical endoscopes, two portable LED light sources and some extra batteries, all resting on a bed of soft, corrugated foam.

When I started in this profession, we struggled to see into the holes we drilled. With only the naked eye and whatever light we could aim down the hole, it was often difficult to get a decent view of a safe's innards. We compensated by drilling big holes, often *many* of them.

Endoscopes were a game changer. No one knows who first adapted medical-grade optics to safecracking, but we owe them a debt of gratitude. These mini-periscopes fit into a small-diameter hole, and through their magical eyepiece we can observe a safe's locking mechanism up close and personal with near-daylight illumination.

Endoscopes — first urethroscopes and nasopharyngoscopes, and later arthroscopes, cystoscopes, bronchoscopes, ureteroscopes, and more — have made my job a dream. When mine aren't in use, they are resting comfortably in the aluminum case I acquired for the sole purpose of scope safety.

It's not unusual for this case to raise an eyebrow. Last week it breezed through security, but not today. The crapshoot is worth the occasional hassle, given the alternative: subjecting forty grand in surgical optics to the tender touch of the baggage handlers.

That is precisely what I had to do for two years following the attacks on 9/11. I was at the airport shortly after commercial flights were allowed back in the air. To my dismay, the luggage screeners (Homeland Security and TSA didn't yet exist) classified my urethroscopes and others as potential stabbing weapons and banished them to the bowels of the plane.

I asked to appeal and was directed to the local representative of the Federal Aviation Administration (FAA), who happened to be at the airport. My modest goal was to carry my scope case onboard like I always had.

The image of someone being stabbed by a slender scope designed to traverse the urethra struck me as hilarious. So rather than pleading my case all serious and somber, I took a lighthearted approach with Mr. FAA and used a little PG-13 language to describe how urologists had once utilized the object he was holding.

Big mistake. Big. Huge! He didn't even crack the beginning of a smile. No reaction at all, other than to blink quickly and put down the urethroscope. I should have told him it was sanitized in an autoclave before being refurbished and sold. That might have mitigated the ick factor and its impact on our exchange.

Humor having bombed, I switched to substance and explained that endoscopes are thin tubes filled with fiber-optic fibers that crumple on impact — hardly the makings of a dangerous weapon. I hoped to see his crewcut move up and down instead of left and right. And I hoped to hear something other than, "I'm sorry, sir, but no." But I didn't. At least he was nice about it.

Out of desperation, I pivoted to what struck me as a knockdown argument. "You know what would be far better for stabbing?" I asked, risking coming off as a smart aleck.

He stared at me and grunted, probably itching to wash his hands. "A sharpened pencil," I answered, trying to appear reasonable rather than triumphant. "For stabbing, a sharpened pencil would be *vastly* superior to a scope." The upshot was straightforward: if pencils were permitted in the cabin, then my scopes should be as well.

That prizewinning point elicited nothing but crossed arms and another firm but polite "no." Mr. FAA's decision was made — I had won the logical battle but lost the war. My scope case was going in the cargo hold, and that was that. Interestingly, pencils have never been banned.

On that trip, I went to Houston while my scopes went to Newark. I found myself in front of arguably the best safe in the world, unable to see in a hole that took me half the day to drill. But it wasn't a complete bust. During the long wait for my scopes to complete their circuitous journey, I reached a personal milestone by reading every word in the *New York Times* without skimming. I even tried (but failed) to finish the cross-word puzzle.

Two years later, the rules were changed to allow scopes onboard again. I was elated, and that feeling has carried me through many subsequent encounters with security. The occasional hassle is a small price to pay for peace of mind, knowing my scopes are safe — as safe as I am at least — in the cabin.

The TSA agent turns my scope case first one way, then another, tilts it up for a closer look at the flush latches, and finally gives up. I watch a frustrated scowl form. "How do you open this thing?"

I know not to touch, so I point a pistol finger at the hinged cover plate Zero Halliburton uses to hide the latch release.[2] "Swing that little plate up and push the button."

She pops it open. Perhaps sensing her earlier abruptness, she half smiles while swabbing the case's interior for explosive residue, then asks if I'm a doctor. Lie and say yes, and I'll most likely be on my way in seconds. But my run-in with Mr. FAA taught me to avoid discussions of scopes and their medical uses, so I opt for the truth. "No ma'am, I'm a safecracker."

It doesn't compute. She points to the scopes. "What? Well, aren't those . . . you're *what?*"

I validate her initial assumption. "Yes, those are used by surgeons. But I use them to open safes and vaults."

The half-smile fades into a quarter before vanishing altogether. *Here we go.* She signals for a second opinion. I hand him a business card and patiently explain how I use the scopes. The quizzical expression tells me he's a tad skeptical.

"Really?" he asks. "You look inside with these, and that opens the vault?"

I nod. "That pretty well sums it up." No point in elaborating. He pauses for a moment, leans over, and wonders aloud, "Have you ever been caught?"

Can't tell if he's serious, so I play along. "Of course not," I deadpan. "Only the hacks get caught." Which is almost (but not entirely) true.

Free to go. Boarding has already started by the time I get to the gate. I lift the case up into the overhead, plop down in the window seat on the pilot's side with the better of my two bad ears facing the aisle and flight attendants, and slide my laptop under the seat in front of me.

So far, so good. No delays or problems of any kind. Everything is right on schedule.

I hear the familiar double chime — we're above ten thousand feet and may now use electronic devices. The flight to Vegas takes a couple of hours, enough time to relax and catch a short movie before walking into the pressure cooker waiting for me at the currency center.

A glance through my digital library reveals mostly oldies. *Meet the Parents, Dumb & Dumber, There's Something about Mary*, a couple of Monty Pythons, and a dozen others I've seen so many times that they merit deletion.

Robert De Niro and Ben Stiller take me most of the way to Sin City. Hmmm, I wonder if it's true that geniuses choose green. Gotta google that one someday. About an hour in, Greg Fokker fumbles his way through grace with a nod to *Godspell*: "We thank you oh sweet, sweet Lord of hosts for the smorgasbord you have so aptly laid at our table this day, and each day, by day, day by day, by day oh dear lord three things we pray to love thee more dearly, to see thee more clearly, to follow thee more nearly, day, by day, by day. Amen."

I can identify with Greg's discomfort, even without the unblinking stare of an ex-CIA agent who gives home polygraphs and trains cats to poop in the toilet. Part of me wonders if there is any difference between prayer and the power of positive thinking. Perhaps one day we'll know.

My favorite parts are over, so I reduce the volume and the player window. The movie becomes background while I walk through final prep for today's opponent.

It's been a couple years since I last battled a Rotary Wedgelock. Add to that the additional pressure of a midnight deadline, and it's especially important that I'm absolutely clear on the step-by-step process. I try to visualize the precise sequence of events that will culminate in success. It's also helpful to highlight foreseeable problems and develop contingency plans.

It's impossible to plan for everything, of course, but I work better when I feel confident, and my confidence rises and falls in proportion to my preparation, especially on difficult openings. When a job goes south, I don't like standing there like a clueless noob, unsure what to do next. This can happen when I don't do my due diligence beforehand.

Meticulous preparation often makes its own luck.

Diebold's Rotary Wedgelock is a thick, beefy vault door. It has a thin, stainless steel fascia plate that I prefer to slide over before drilling. That way, the stainless can be slid back into place after I repair the hole, leaving no evidence that the vault door was ever penetrated.

But both dials and the handle must be removed before the stainless plate can slide sideways. This can be a time-consuming pain in the buttocks, and sometimes the stainless just won't budge. On second thought, maybe I should leave the stainless in place and drill through it to save time. The repair site can be covered with a manufacturer's decal; the public never sees this door anyway. *Good idea.*

The 737 begins its slow descent while the movie credits are rolling. Even though I've seen it many times before, I chuckle silently when Greg Fokker stares sternly into the nanny cam and spins into a karate spaz. Still smiling, I stow the laptop, take out my phone, and shuffle one of my favorite mellow playlists.

The plane kisses the runway so softly that I barely notice. Tip of the hat to the pilot. I feel good. This is going to be a textbook vault opening with no glitches. Murphy is going to stay home.

At baggage claim, I spot the case containing my portable drill press floating my way on the conveyor belt. I grab it and look over to the plastic curtain, expecting my Beemer case to pop through.

I wait and wait, but Beemer doesn't appear. I text Hitch. He calls. "Problem, eh?"

"Yeah, my main tool case hasn't shown up."

"Should I come in?"

"Nah, I can handle it."

"Well, I need to move, so I'll park in the garage and text you where."

"Good enough, Hitch. See you in a bit."

This is starting to feel like one of my Phoenix jobs. Flew there a few years ago to open a vault for the Desert Energy Credit Union. The vice president and branch manager met me at the airport, but neither of my cases arrived. I hadn't a single tool or drill bit to work with.

They took me to Home Depot, where I spent four hundred smackeroos on a drill motor and a pile of drill bits ill-suited for the task. With no drilling rig of any kind, I spent the next few hours in utter agony, trying to brute force a hole in a vault door that just laughed at me.

Fortunately, the airline found my bags and delivered them a little after midnight. I was never so happy to see my equipment. With the proper tools and drill bits, I finished the job in about two hours.

If Vegas turns into Phoenix, I'm toast.

All the other passengers have retrieved their bags and departed. *I can't believe this.* Without my custom tools, I won't get this vault open before midnight. I'm starting to feel a little queasy.

I've taken two steps toward customer service when the conveyor belt starts up again. I look back just as my Beemer case pushes through the plastic curtain and onto the conveyor belt in front of me. I know it's mine from the ragged red ribbon that's been around the handle for as long as I can remember.

Whew! What a relief.

CHAPTER FOUR

Hitch and the Vault

WHEELING THE CASES THROUGH THE PARKING GARAGE AT MCCARRAN, my cold sweat slowly begins to evaporate. I spot the white Diebold van, load the cases in the back, hop in the passenger seat, and shake Hitch's outstretched hand.

"Good to see you, Hitch. It's been a while."

"Nice to see you too, Dave," he says, putting the van in reverse. "Glad all your tools finally arrived. Hey, do you really think we can pull this off?"

I nod. "I think so," clicking on my cell and glancing at the time. It's eight-thirty. "Gonna be close, but we can do it."

Hitch and I have worked together on at least a half dozen vault jobs in the Silver State over the past three or four years. Soft-spoken, dependable, and easy to work with, he resembles a bespectacled Michael Keaton, but with a bit more hair.

Hitch cuts his headlights as we pull into the parking lot. Millions of dollars flow through this facility, but you'd never know it from the outside.[1] The unmarked building stands in a row of unmarked buildings in an ordinary-looking industrial park. It's protected by many layers of security, some of them invisible to the naked eye.

No time to dawdle, we grab the gear and head to the door, which leads directly into the mantrap — a small room with a door on each end, only one of which can be open at a time. When the buzzer sounds, we open the first door and enter the trap room. The door closes and we are temporarily imprisoned, unable to go forward or back.

Security guards surveil us through bulletproof glass and ask a few questions through the intercom. We pass the admissions exam and are buzzed through the second door, where we are met by more security personnel who hand us scarlet aprons. These identify us as visitors and must be worn for the duration. I tie Hitch's apron behind him and he ties mine.

There are video cameras and armed guards in every direction. To my knowledge, no big bank currency center anywhere in the country has been successfully robbed. It's easy to see why.

We follow our escort down a hallway and enter a large room, where I spot the vault door along the far wall. I was told our opponent is a Rotary Wedgelock, a thick but fairly easy door to penetrate. But this is no Rotary Wedgelock. This is a Pacesetter, a much thinner door that is far more difficult to drill.

I can see how the mistake was made. Both doors are Diebold products with two dials and four-spoked handles that exhibit a sibling similarity. But they can be distinguished by the locations of the dials and a few cosmetic differences.

I have drilled just about every make and model of bank vault, most of them many times. The Pacesetter is a tough door with an extremely tough layer of hardplate. You won't find the word *hardplate* in the dictionary, but it exists as a term of art with a long pedigree in the lexicon of the safe and vault industry. It's a portmanteau, combining "hardened" and "plate" into a single word that means exactly what it sounds like. There are many different hardplates, but they all have the same primary purpose: to deter drill bits.

I have two different drilling methods that work well on the hardplate in this vault. One requires a few diamond bits, the other a few dozen carbide bits. But we have a problem. The bank bag containing my diamond bits is on the workbench at home and so is the bag filled with short carbide bits.

The Pacesetter hardplate is notorious for dulling a carbide bit's leading edge, so I use each one for only a few minutes and then replace it. Rinse and repeat until the hole is finished. This is both a show of overwhelming force and a war of attrition, where quantity becomes a quality, and the path to victory is as simple as sending more drill bits into battle than the opposition can withstand.

Because of the pressure on the bit, its shank cannot protrude very far outside the hole without bending or breaking. All those long bits with full-length flutes are now useless. I need *short* bits. I packed two. And you can't buy what I need at any hardware store. These are special bits, sharpened to a razor's edge on a diamond wheel.

Not good, not good.

Hearing me grouse, Hitch puts down the tool bag. He says he has a few short bits in his van and heads out of the facility to retrieve them.

I unzip my Beemer case next to the vault, exposing a few tools of my trade. And I do mean a *few*. Having to make do with a skeleton set of tools is the risky part of flying around the country. I've done it hundreds of times with zero failures, but one of these days I'm going to find myself missing something crucial to success on a particular job — like short carbide bits or diamonds for a Pacesetter.

As Hitch approaches, carrying a small box, I glance at my cell. It's almost nine o'clock. We have three hours to get this vault open and skedaddle. Or else. *Stop checking the clock and get to work.*

Hitch's box contains three short, carbide-tipped 5/16″ bits. Together we have a grand total of five. *Five.* My entry method requires two or three *dozen.* I feel a twinge of panic. My mind is racing. By gawd, this is *not* going to be my Waterloo. Think, *think.* With zero diamonds and only five short carbides, we need an alternative method.

The only thing that comes to mind is putting enormous pressure on the drill motor, more than I have ever used before. My portable drill press isn't suitable for this task. It'll pop off the vault door under extreme duress.

Fortunately, I always bring my lever rig — the simplest and most versatile drilling rig a safecracker can own. It's basically a long bar that applies pressure behind the drill motor. Custom-made from aircraft aluminum, mine telescopes out to nearly three feet in length and weighs less than three pounds.

I don't know whether the high-pressure idea will work, but I'm relieved to have a rational plan. If it succeeds, great — this old dog will have learned a new trick. And if it doesn't, well, we'll cross that bridge when we get to it.

I remove the lever rig, the Fein drill motor, and a high-speed bit from the case and lay them in front of the vault door. Hitch grabs the extension cord and searches for an outlet while I look up the drill point, carefully measure it out, and mark it with a black Sharpie dot.

Safecrackers make Sharpie dots for the same reason plastic surgeons draw Sharpie lines on a patient's body — to create a template for penetration. I use a drill bit rather than a scalpel, and the dot pinpoints the precise location for making the hole. Consequently, on doors that are difficult to drill, I spend extra time measuring, remeasuring, and re-remeasuring before firing up my drill motor.

If I had a motto it would be: *measure thrice, drill once.*

I remeasure the drill point over and up and discover that my Sharpie dot is a hair too high. So I make it a little larger, adding all the new ink to the bottom of the dot. It looks good.

If I were a smoker, I'd excuse myself at this point for a two-puff break. But I'm not, so I take out my phone and open the Words With Friends game my wife and I are playing. I don't have any vowels, so I put an "S" on the end of QUOTE for a meager sixteen points. Valerie is eighty points ahead. I have to step up my game.

The point of this short break isn't to indulge my WWF addiction, but to clear my mind before *triple*-checking the location of that all-important Sharpie dot. Any thirty-second distraction will do.

I have opened somewhere around ten thousand safes and vaults, and the difficult ones still give me butterflies. This is a good thing. The tiniest detail can make all the difference, so allowing myself to drift on autopilot isn't an option. Complacency degrades competency, and the most effective inoculation is a nagging, low-level fear of failure. If I recognize it and adopt strategies to prevent it from morphing into brain freeze, this fear will function like a double-shot espresso rush — heightening sensory awareness, sharpening the cognitive faculties, and preventing me from glossing over crucial details. So long as I continue feeling butterflies on tough jobs, I'll continue triple-checking my drill points.

I measure out the location of my Sharpie dot one last time, and it's right on the money. We're ready to rumble.

I depress the trigger on the Fein and push the 5/16″ bit through the Sharpie dot into the stainless steel. Shavings begin to flow as the drill bit digs into the door, and my shoulders relax a little. There's a certain comfort in the familiar, and few things are more familiar to me than this.

After a few inches of penetration, the speed of my drill motor increases and the shavings stop coming. We're at the hardplate. I confirm by inserting a straight-view scope in the hole for a close look. No doubt about it. We're looking at a thick piece of Impervium. This name wasn't dreamt up by the writers at Marvel; it's the actual, trademarked term for the nasty hardplate Diebold laminated into these doors. Impervium provides extraordinary resistance to a variety of attacks, including and especially, drilling.

It's time to test the high-pressure theory. Chucking up the first 5/16″ carbide bit, I look over at Hitch. "Ready?"

He smiles and tilts his head forward. Hitch is up for whatever — I like that. He's on the end of the bar for maximum leverage as I hold the drill motor straight. We don't want to veer off target. An experienced driller himself, Hitch knows not to lunge forward but to smoothly and gradually apply pressure. He gently leans into the bar, pushing the drill motor forward, and the carbide bit makes contact with the Impervium.

Nothing.

I glance over. "More, Hitch."

He pushes a little harder. Nothing.

"More."

He pushes a lot harder. Still nothing.

"Come on, Hitch, put some *ass* into it."

His eyes narrow, and his mouth forms a straight, pursed-lip grin. He leans into the bar with just about everything he has, and I can barely hold the motor straight. The pressure is so great that I'm concerned the shank might break or the tip might snap off. If that happens, it will significantly increase the chances of our missing the deadline. Not the desired outcome.

Hitch's glasses slide down a little as sweat begins to form on his nose. He's working hard, and it's now starting to pay off. The sound of the motor changes slightly, and a few tiny slivers of ultra-hardened steel roll out of the hole and fall onto the shop rags I taped to the floor.

The slivers keep coming, but Hitch needs a break. I hate to stop while we're making progress, however slight, but what he's doing is absolutely exhausting. It's a good time to check the tip of the bit. The carbide insert is intact, but its leading edge is now dull.

I chuck up the second bit and we start again. Hitch leans harder and harder into the bar, very near his max, until little slivers start to come. He maintains steady, heavy pressure for a couple of minutes, and then we trade places. Hitch holds the motor while I push on the leverage bar.

Holy moly, I much prefer being the holder rather than the pusher. This is like doing a bench press and holding the weights at full extension until your arms almost give out on their own. It's grueling, grunt labor, but it's working faster than any method I have ever used on Impervium.

Back and forth we go for about forty minutes. We're making excellent progress, but the clock is ticking and the drill bit situation has me concerned. Hitch and I are both wringing wet as I reach into the box and pull out the fifth and last bit. Agnostic I may be, but I mutter a silent prayer to the goddess of safecracking. If this bit doesn't get us through, we'll have to consider alternatives, all of them ugly and time-consuming.

I vow that this will never happen again. It was foolish of me to leave my diamond bits at home and to bring only two short carbide bits. Thank goodness Hitch had a few. In the future, I simply *must* see a photo of my opponent before I pack tools and jump on a plane.

Sometimes, though, not even a photo is foolproof. Recently I dealt with an overwind on one of the two vaults at a jewelry wholesaler in Seattle. I asked for and received a photo via email. It was a simple model that I can open quickly by inserting a small, robotic arm through the ventilator hole. (All modern bank vaults are required to have such a hole, to provide air in case someone is trapped inside.) No holes need to be drilled. I took my sons with me on this "easy" job, only to arrive onsite and find an entirely different make and model.

The person who took the photo assumed the two vaults were identical, so he saved himself the hundred-foot walk to the one they needed me to open, and snapped a photo of the closer vault instead. They look similar from the outside but are completely different internally. My boys were disappointed at not getting to see the robotic arm in action, and I was reminded that there are no guarantees.

The drill motor, very warm to the touch now, changes its tune slightly and I motion for Hitch to ease up on the pressure. The bit is most vulnerable to snapping off while breaking through the hardplate. The stress is no longer evenly distributed across the carbide tip, and any side-to-side movement at all heightens the risk of breakage. During these crucial seconds, safecrackers hold their motor (and their breath) as steady as humanly possible.

I don't let my Fein move an iota left or right or up or down, and it jumps forward a little as the bit breaks through into the air gap beyond. I look over at Hitch, who is sporting a knowing smile. There is now a 5/16″ hole through the Impervium, and in record time. We made it.

I chuck up one of the footlong, high-speed bits and drill through the remaining mild steel, then the mounting bridge, and stop as the bit touches what I hope is the timelock case. We have time to spare, so I remove the bit and peer down the drilled hole with a right-angle scope. I see the thick rubber grommets that cushion the timelock against the door. This confirms that our hole has hit the target right on the money. Excellent. I put the long drill bit back in the hole, drill through the timelock case, and stop.

I hear footsteps and glance over my shoulder. It's the site manager. Smiling professionally, she asks, "How're we doing? Getting close?"

"We are," I reply. "Should be just a few more minutes." My deadline is hers, too.

The scope goes back in the hole, along with a special wire I use to defeat certain timelocks. I watch the wire enter a void inside the timelock. I can see that the bend on the tip is a little too short, so I pull it out, rebend it, and try again. The bend contacts the blocker inside the timelock and pushes it down, out of the way.

I look over at Hitch, who is waiting with his hand on the handle. He and I have been here many times before. I nod without saying a word, and Hitch rotates the handle to unlock the vault.

I roll up the shop rags, trapping the drill shavings inside. This simpli-fies cleanup. I put my scopes back in their case, move the tools to the side, and watch as Hitch swings open the three-ton door. (In cases like this, where the client already knows what's inside, there is no "reveal," so we go ahead and pull the door open. But when the contents are a mystery, we let the client do the honors.)

The manager and her team rush over. "Fantastic," she exclaims, and they enter the vault to take care of business. Through the open door, I see the tables and currency sorting machines. Soon there will be paper money flying in every direction. We got the vault open with an hour to spare, but still, they're scurrying to get the money train chugging on down the track.

Tonight's job was the result of human error. Too much time was wound on the timelock. The vault would have opened on its own, but not until tomorrow, which wasn't acceptable. So the bank called Diebold, and Diebold called me.

It's a good gig.

CHAPTER FIVE

The Italian Job

CHARLIZE ISN'T DOING IT RIGHT. SHE HAS THE MOTOR TURNING WAY TOO fast for the material she's trying to penetrate, and she's swaying side to side, about to snap off the drill bit. Good grief, she's behaving like a novice!

We're on the set of *The Italian Job*, a caper flick with Ms. Theron cast as a professional safecracker. Her co-star, Mark Wahlberg, is waiting anxiously on the sidelines, but Charlize is my first priority. I ease in from behind like a golf instructor, gently lay my hands on hers, and whisper commands in her ear. She slows down, stops swaying, and turns her head to catch my nod of approval. Oh my, she smells good. *But do I?* Oh man, I hope my breath doesn't reek. Should've popped a Tic Tac.

My nostrils flare, and I wonder how golf instructors do their job without getting slapped. I feel my face and neck flush, and back up a little so that physical contact is made only with the hands. But I'm still uncomfortable. Massively so.

My mind feels numb, and odd sensations begin infiltrating my consciousness. My eyes slowly open. Charlize is gone. I'm in bed, alone, and my heart is pounding in my ears.

Wow, that was wild.

I'm as honest as the day is long, but name the caper flick, and in my dreams I've stolen a scene or two in it. These mental movies are usually all business, with me playing the hero who steps in to upstage Clint Eastwood, Robert De Niro, or James Caan and crack the uncrackable safe.

This one was different. To my memory, this is the first time one of my safecracking fantasies has included any spicy innuendo. My wife and

I share particularly intense dreams and nightmares, and Valerie will find this one amusing. I've remarked several times that she could be Charlize's older, shorter sister.

I glance foggily at the nightstand. One of the first things I do in a hotel room is angle the clock toward the bed so I can see it without moving a muscle. That way, if I become semiconscious in the middle of the night (as often happens), I can check the time with minimal effort and try to get back to sleep.

Staring at the blurry numbers until they come into focus, I see that it's three-twenty. Lovely. Waking up well before sunrise is one of my least favorite things about late middle age. I hope it doesn't worsen as I approach the elderly years.

I'd like to rest for a while before heading to McCarran to catch my early-bird flight home. But the cobwebs are clearing quickly, against my will. Like it or not, I'm waking all the way up. Better get coffee and shake off this latest REM adventure.

I don't have a clue why I constantly have these dreams, and a psychoanalyst would likely have a field day with my subconscious meanderings. I probably don't want to know what's really going on upstairs. But there is a connection in this case — Charlize, Mark, and I almost crossed paths.

I missed a golden opportunity to sell myself as a consultant on *The Italian Job*. The year I skipped my industry's national convention was the year the movie's producers showed up at my trade publisher's booth and bought a bunch of my safecracking books. Had I been there, I would have done my best to wow them, woo them, persuade them that *The Italian Job* absolutely, positively, without a doubt, needed my expertise.

I didn't find out about it until the following year's convention when my editor offhandedly commented on the movie people who bought my books.

"Uh, what?" I stammered. "I wasn't here last year. What movie?"

It was too late. *The Italian Job* was well into production by then. And what a production it turned out to be. The working pros in my field love movies that showcase safes and vaults, and *The Italian Job* is one of the best. Even the silly and implausible safecracking scenes play beautifully onscreen. Still, it would have been great fun to introduce Valerie to Charlize.

It's on my bucket list.

TUESDAY

CHAPTER SIX

Good Morning

I'M IN THE NEIGHBORHOOD, ALMOST HOME FROM THE AIRPORT, WHEN my phone rings. It's a kind sounding lady who was referred to me by a locksmith friend up in Salmon Creek. He refers safe jobs to me, and I refer car and key jobs to him.

The combination to the gun safe in the lady's garage has been lost, and she has a few questions for me.

Can it be opened without causing visible damage? "Yes. As long as all the parts are there and functional, I will either manipulate the safe open or drill a tiny hole that will be repaired and invisible when the job is complete."

Can the job be done between noon and three o'clock today? "Yes. In most cases three hours is plenty of time, and my afternoon is wide open."

She doubles back to verify that no one will be able to tell the safe was opened, that I will be done and gone by three o'clock, and a red flag goes up — she hasn't even asked for a price yet.

"Ma'am," I begin, as the van rolls into the driveway, "do you mind telling me why the timing is so important? I don't mean to be nosy, but for liability reasons I will need to verify ownership of the safe."

It spills out. She suspects infidelity and wants to look inside the safe for evidence. I don't ask what she expects to find. Her husband gets off work at three o'clock and is usually home within minutes, but sometimes not for hours. She doesn't want him to know about me.

Opening a safe for one side of a marital dispute is an occupational hazard I prefer to avoid. The atmosphere is always negative, often charged, and sometimes dangerous. Safecrackers have been violently assaulted and even killed in the line of duty.

We occasionally find ourselves standing in for therapists or confidantes. I'm neither but want to convey to this poor woman that her idea has no upside. If her husband *is* cheating, there are much more reliable ways of finding out. And if he isn't, having me crack his safe effectively sets her up as an untrusting partner. This can lead to feelings of guilt, which may result in her confessing the break-in to hubby, thereby causing *him* to feel violated. Like falling dominoes, negative synergy in a marriage can be difficult to stop once it gets started.

None of this is any of my business, so I don't approach it directly. Instead, I stall, and ask her to email me so I can send over the paperwork to sign. As food for thought, I suggest that she consider asking him for the combination. In our state, the safe's contents are half hers.

She says she'll think things through and get back to me. We'll see. In most of the messy relationship cases I push back on, I don't hear back. I hope that's a good thing.

I turn off the van just as a text comes in from Harry's Locksmith Service, another local client of mine, asking for a bid on opening a gun safe. Wondering if it's the same one, I text back, *Does this job have to be done between noon and three, leaving no visible evidence of entry?*

Just as I press SEND, a photo appears. It's not the same one. The lady's gun safe has a dial and a mechanical lock; the safe in Harry's photo has a keypad and an electronic lock. The code is known, but the e-lock is malfunctioning and the safe won't open. Harry's techs were onsite yesterday and tried brand new batteries, to no avail, so it probably needs to be drilled. I text a price, and we agree to meet a week from tomorrow, after the owners have returned home from a weeklong trip they just embarked on — without their favorite rifles.

Safe owners sometimes ask if electronic locks malfunction at a higher rate than their mechanical counterparts. They do, but there are compelling trade-offs that make e-locks attractive. They offer multiple user codes with auditing capabilities, single or dual custody, time delay, and more. What persuades most of us, though, aren't the fancy bells and whistles, but how fantastically easy most e-locks are to operate.

The difference is night and day. Mechanical locks have a complex opening process — four turns this way, three turns that way, and so on — and they are unforgiving of even the smallest mistake. Misdial any of the numbers by a tiny fraction and the lock won't open. You have to start over and dial the combination again. It's aggravating.

By contrast, most e-locks simply require you to enter six digits on a keypad. In seconds your safe is open. Because e-locks offer user-friendly features that mechanical locks lack, they are here to stay. That's all well and good, but there is a troubling fact you should know: many manufacturers program an override code into the e-locks they put on their safes, and in most cases they are keeping this information *without the buyer's knowledge or consent.*

What does this mean? It means there may be an electronic backdoor into your safe. No, I'm not kidding, and here is the kicker: no matter how many times you change your *user* code, the secret *override* code will still be there, like a Manchurian candidate awaiting activation.

Why do manufacturers do this? For the same reason they record the original combinations to the safes and vaults they sell: as a convenience for the buyers of their products, and to protect their investment during shipping. There are times when the original combo or the override code can save the day and open the container without drilling. This is handy, of course, but it's disturbing that most buyers have no idea that the single most sensitive piece of information about their safe is kept on file in a factory somewhere.

My advice: if you don't want a manufacturer to know your original combo, have it changed by a reputable locksmith or safe technician. And if you are a DIYer inclined to change your own combo, please do it with the safe door *locked open.* That way a mistake won't result in an expensive lockout.

Regarding the possibility of an override code in an e-lock, my advice again is to bring in a professional. They can identify your lock's make and model, determine whether an override is present, and change it to something you prefer — or, in many cases, delete the backdoor altogether.

I head into the house and make a beeline for my computer. With no jobs on today's docket, it's time to write. My safecracker magazine, due next Monday, is only partly done. Okey dokey. Nothing spurs productivity like a looming deadline.

For two decades, I have penned a quarterly trade journal dedicated to the fine and sometimes not-so-fine points of professional safecracking. *The International Safecracker* is a labor of love, but I'm always scrambling to get the darn thing done on time.

I admire writers who can crank it out effortlessly. William F. Buckley famously typed op-ed pieces (on a typewriter!) during his daily limo ride to work, and entire books during his annual month-long vacation on the slopes in Switzerland. Most of us can only dream of such output.

I need to decide whether to dedicate the entire issue to a narrow topic examined in great detail or to cover several topics superficially. A classic breadth versus depth dilemma. The former is easy — just build the issue around a few recent jobs. I can knock that out in a week or two. The latter takes twice the time, due to the research required, but is more valued by the industry pros who read my ramblings.

Fingers slow dancing on the keys, I'm midway through the first draft of an article about a malfunctioning vault door on a panic room when my cell phone begins to vibrate against the well-worn oak desktop. It's Alberto from Loomis Armored Car. They're locked out of an ATM at the Arlene Schnitzer Concert Hall in downtown Portland and need it opened ASAP.

I haven't been to "The Schnitz" since the band days. We opened there for Emmylou Harris, where I had a memorable experience a few hours before showtime. I came in for sound checks and rounded the corner toward the dressing rooms just as Emmylou came around the same corner from the opposite direction. We both had guitars strapped on, and the head of my Telecaster swung within an inch of her jumbo acoustic.

"Whoa, excuse me!" I exclaimed, as we stopped dead in our tracks and leaned away from each other. This was *not* the way I wanted to meet my all-time favorite female country singer, but Emmylou was gracious. "That was close!" she replied, unfazed and smiling, continuing on her way to the stage.

Despite this inauspicious start, the evening went well. The crowd responded enthusiastically to our opening set, and we got a nice mention in the *Oregonian* the next day, when the music reviewer listed among the evening's highlights my "blistering country lead guitar" and our bass player's poignant cover of David Allan Coe's classic hit, "The Ride." The reviewer also had some very nice things to say about the headliner, so maybe, just maybe, Emmylou read the article and forgot all about our earlier encounter. Oh, the audacity of hope.

I planned to spend today working on the magazine and dressed accordingly. So I grumble a little while swapping out the sweatpants and tennis shoes for my standard work uniform: tan Dockers, white t-shirt, black vest, and black fisherman sandals. Yeah, sandals. But no ponytail, at least not yet.

Conservative hip is a work in progress.

Chapter Seven

Alberto and the ATM

Downtown parking in just about any big city is a hassle, and Portland is no exception. But today I luck out and squeeze into a tight spot only a block from the job. I grab the Beemer case and wheel it down the sidewalk to meet Alberto in the lobby of The Schnitz.

I enjoy cracking safes with Alberto. Big and barrel-chested, he wears a perpetual smile on his clean-shaven face and knows how to run interference without ruffling feathers. His holstered Sig Sauer P220 keeps most passersby at arm's length — a nice perk when the job is out in the open.

We walk inside the brick building. A few feet down from the ticket counter, next to a green courtesy phone hanging from the wall, stands a tall, narrow, blue and gray ATM from South Korea. It looks familiar. An image of Jesse Pinkman swinging a sledgehammer pops into my mind and I smile in recognition.

My South Korean opponent is the same make and model as the one stolen by a hapless meth-head couple in the ATM episode of *Breaking Bad*.[1] Jesse helps them pound on it but gets knocked senseless. On the ground, he groggily watches one meth-head taunt the other by calling her "skank" repeatedly, mercilessly, and eventually to her breaking point.[2] Jesse stumbles to his feet just as she tips over the ATM to crush her tormentor's skull. It's a creepy but memorable scene.

I have a trick that will swing open the money door in just a few minutes, using a tiny drill rather than Jesse's inelegant sledgehammer. But there is a negative to contend with: we are in full view of the people coming in to buy tickets.

Safecrackers dread safes that are exposed to the public. Our unanimous preference is to work in a back room, away from prying eyes and wannabe comics. Heavily trafficked storefronts are the absolute worst. You're completely exposed to armed robbery once the safe is open. You have to deal with the folks who can't resist the urge to tell you how to do your job or trot out an entirely predictable joke. Worse yet, you have to be constantly on guard for unsupervised children who may wander into harm's way. Consequently, these jobs require a little more setup and a lot more vigilance.

I run an extension cord from the wall outlet to the ATM, and tape it to the carpet. I lay my tools down in a half circle about six feet in front of the ATM — a barricade most people won't cross. Plus I have Alberto to field questions and collar anyone who attempts to breach my perimeter.

Sure enough, before I even fire up the drill motor: "What's going on? You guys breaking into the ATM?"

My back is to the questioner, but I hear Alberto answer, "Yeah, it broke."

"Gonna git the dynamite?"

Just once, I'd like to respond to this frequently asked question by pulling out a (fake) stick of dynamite, duct-taping it to the safe, and lighting the (real) fuse. Mr. Witty's reaction would be epic, especially if he were to see me bolt for the door. But I fear he would have a heart attack or slip and fall during his frantic escape and break something. For that reason, I'll probably never do it, but the long-closeted prankster in me smiles every time the idea crosses my mind.

Alberto chuckles at the question. "Nah, no need for the boom. We can drill this one." Curiosity satisfied, Mr. Witty takes his place in the ticket line. I'm marking the drill point when he is replaced by Mr. Wittier, who informs us that C-4 or plastique will do the job nicely. Content to let Alberto do the talking, I briefly turn and smile mutely.

But Mr. Wittier wants a reply from the guy who is about to dirty his fingernails. "Hey," he half shouts, "you gonna blow it?"

"No sir," I reply, impassively. "The fire marshal won't sign off on that. We have to crack it the old-fashioned way."

"And what way is that?" he persists, still not satisfied. Alberto casually takes a step in his direction and offers a classic, delivered with a smile and wink. "We could tell ya, but then we'd have to kill ya."

No sooner has Mr. Wittier disappeared up the stairs and out the door, when he is replaced by Mr. Wittiest, who mentions MacGyver and sanded fingertips. And so it goes, for the duration of this and every other safecracking job in plain sight.

The reason I'm here is that the KABA Cencon electronic lock has failed. No surprise there. More than ninety percent of the ATM lockouts I deal with are caused by a malfunctioning Cencon. I love this little rascal!

Unlike most other electronic locks, the Cencon has no back door, which makes it one of the more secure e-locks on the market.[3] Because of the way it works, it's used almost exclusively on ATMs and other containers involved in cash handling. Service personnel, such as armored car drivers and cash replenishers, must contact a central dispatcher to receive a "one-time code" whenever they need access. Personnel must also present their unique electronic key fob to the lock before entry, providing an additional layer of security and accountability. These features of the Cencon have reduced insider theft, once the bane of the ATM world, to nearly nothing.

My preferred opening method for the Cencon requires only a tiny 1/8" hole to be drilled through the ATM's door. I'm about halfway through the steel when I look over my shoulder and spot Alberto in veiled distress. The edges of his mouth are still turned upward, but his furrowed brow strongly indicates he would like Mr. Wittiest's monologue to end, sooner rather than later.

It's my turn to run a little interference. I stop drilling and pop both battery doors outward to cut power to my hearing aids. The ability to turn the world down when it gets too loud is the best part of being hearing-impaired. I flip my Fein motor into hammer-drill mode and start making some serious racket. It sounds like a mini jackhammer gone wild. Our antagonist covers his ears and beats feet for the exit. I wink at Alberto, who gives me a big grin and a thumbs-up.

I slide my battery doors closed and wait for my bionic ears to boot up. "Sorry folks," I announce to the people in line, trying not to look guilty. "The loud part is over — we're almost done."

I feel the double vibration in my upper vest pocket for the umpteenth time since arriving onsite. I hope the people texting me don't need immediate replies.

The now-dull bit is replaced with a new one, and I finish drilling the tiny hole through the door and into the failed lock. It's time to inspect the hole, so I open my pistol case. There is no gun inside — I'm not going to blast my way in. Besides, even if I did own a gun, it would be in an ankle holster for easy access, not in a storage case where it would be useless against bad guys who prefer the element of surprise and don't give timeouts.

I like the pistol case because it fits inside my Beemer case, and it contains the scopes I need to do most of the easy jobs I do locally. Inside, resting side by side on the soft, corrugated foam, are four of my favorite arthroscopes, an LED light source, and a few extra batteries. If I need longer scopes, I can always run to my van and grab my aluminum briefcase with the footlong cystoscopes or my large Storz case with the two-footers and flexible six-footers.

The 4mm scopes won't fit in a 1/8" hole, so I reach past them for a 2.7mm pediatric arthroscope and insert it for a look inside to confirm that the hole is where it needs to be. It is. Then I insert a long, custom-ground 3/32" pin punch into the lock, gently tap the lever out of the slide, turn the spindle, and open the lock.

"It's ready, Alberto."

He ambles over, turns the handle, and swings open the ATM door. Inside is a single cassette that's nearly empty. A full cassette holds two thousand bills, which amounts to forty large when those bills are twenties. Many ATMs have three or four cassettes. Do the math and you'll understand why I appreciate having an armed guard onsite. (As I was writing this book, a Diebold technician was shot to death while working on an ATM in Cincinnati. Guards are no longer a luxury in high-crime areas. They're mandatory.)

The hole is patched with a 1/8" bearing ball and a little steel epoxy, which is carefully sanded and blended to cover the repair site. A new Cencon electronic safe lock is installed, and we're done.

Alberto and I shake hands and bid each other farewell. He's still smiling as he waves and turns toward his vehicle. Our world could use a few more infectiously upbeat souls like him.

Chapter Eight

Temptation

On the walk back to the van, I scan a few of the texts that came in while I was drilling the ATM. The only job-related one is from Perrill Smith, a Diebold technician on the Portland-Metro team. Perrill is working on a night depository that suffered a burglary attack last night. The keypad was hammered into oblivion. He's hoping to get it open with a replacement keypad but is checking my availability in case that doesn't do the trick.

I text him that I'm in the area and can head his way immediately. I'm steering the van into traffic when he texts me back: *Stand down, Dave. It opened right up.*

The amateur attack must not have damaged the cable that comes out of the spindle hole and connects to the night drop's keypad. I'm glad Perrill got it open and the bad guys didn't.

I've been asked many times whether I'm ever tempted by the allure of a big score to break into somebody's safe. The answer is no. I couldn't do it. This doesn't require will power on my part, thank goodness, because there is no felt temptation to overcome.

And it's not just me. My colleagues around the world are of the highest integrity and express the same sentiment. Many of us have discussed the riches that could quickly be acquired if we were so inclined. We have had these talks over enough drinks that I can say with a high degree of confidence that the public is safe.

Though making a profit is necessary and nice, we're not in this game for the money. We're in it for the continual call to adventure and the challenge, which is all about solving the puzzle and opening the safe, not ransacking its contents. Most of the time, we don't even notice what's inside, unless a client's reaction is so strong as to momentarily draw our attention.

Among the safecracking elite, it takes some pretty wild hypotheticals to generate even a reluctant "hmmm, that's interesting." Here is one: imagine a convicted serial killer serving life without parole, with a hidden safe containing a million dollars. Further imagine that you know with *Minority Report* certitude that his house is going to suffer an electrical fire and burn to the ground, turning the money into ashes inside that low-end, non-fire-resistant safe.

In other words, the money is owned by a bad guy who will never again have access, and it's about to be destroyed by an accidental fire. Question: if you had the opportunity to open his safe and take the money five minutes before the first flame licks the curtains, would you? Assuming your answer is no, let's tweak the hypothetical. Instead of taking the money for yourself, imagine that you could distribute it among the families of the serial killer's victims as a secret benefactor. Would you do it then? This version has given a few of my friends pause, but none of them have said yes.

To get an actual, "Okay, I'd do it," requires an extreme hypothetical, such as: *break into the safe or the entire universe will implode in two hours!* Yeah, whiskey-laden conversations among professional safecrackers can be all kinds of fun.

One of the few things I require of a client is that they never leave me alone with valuables inside an open safe. You might wonder why I insist on this, given that I just said I would never take anything (except to save the cosmos, of course). The answer is to avoid being *accused* of stealing something. Yes, it happened. Once.

My very first job, as a scrawny sophomore, was cleaning the key machines and sweeping the brass filings off the floor for a second-generation locksmith named Eugene Corey. We worked on a lot of safes and vaults together, but the most valuable lessons I learned weren't technical, they were about posture and protocol.

Mr. Corey and I went out one afternoon and changed the combination on a home safe. When we got back to the shop, we received a nasty phone call from the husband, who insisted we had taken a jewelry box from the top shelf inside the safe. If we didn't return it immediately, he was going to call the police.

It was unnerving to be accused of something we didn't do, and we realized the difficulty in proving a negative. Fortunately, the guy called back a few minutes later and apologized, after learning that his wife had moved the jewelry box that morning, prior to our arrival.

Rather than jump on the husband for jumping the gun, Mr. Corey, as always, was gracious. "Well, I'm just glad you found the missing items. Thank you for putting our minds at ease."

He hung up the phone and turned to me. "We just learned a good lesson, son."

That lesson became my third commandment of safe and vault work: **Never be left alone with an open safe that contains valuables.**

Nearly all attempted safe burglaries are committed by dunderheads who do the darndest things, like cutting off a safe's hinges and breaking the handle or hammering on the dial or keypad. Brute force attacks may provide temporary relief from testosterone overload, but they lock up the safe even tighter. The few burglars who *do* get the goodies are typically attacking cheap tin cans that hardly deserve to be called safes.

There are exceptions, however. While it's extraordinarily rare for a pro to turn amateur (aka criminal), it has happened. In all the years I've been running my club for safecrackers, two members have been ejected following their convictions related to safe burglary.

Mike Ornelas ran his own safe company in Central Florida. Soft-spoken and likable, Mike was a working professional, trusted with safe-cracking secrets in all but the innermost circle. He attended many of my classes and was once mentioned in an article in my club's magazine.

I woke up one morning to the news that Mike had been arrested along with a police officer and a private investigator. They had committed a string of safe burglaries at jewelry stores and supermarkets in the southeastern United States. The investigation widened and snared a jeweler and a fence, too.

They got caught when they put stolen jewelry on eBay. Seriously. A unique pocket watch was spotted by the jeweler who sold it. Unbelievable.

Mike and the other defendants turned state's evidence against the police officer, who received six years in federal prison. Mike and the rest of the gang received sentences ranging from twelve to twenty-four months. The prosecutors were understandably after the guy with a badge.[1]

The other time was an even bigger shock. British safecracker Roy Saunders had long maintained a high profile in the industry. A slight but elegant man in both appearance and demeanor, Roy was a master raconteur who could hold an audience's attention for as long as he wished — in a class, on a bus, or at a bar. He once toured the States, demonstrating the burglary resistance of high-security safes, mostly to jewelers and insurance executives. Everyone in the business knew Roy or at least knew *of* him.

Again, I woke to a slew of messages from around the world, all bringing impossible news: Roy Saunders and a gang of thieves had been arrested while attempting to burgle Cartier in downtown London. My first reaction was denial. *Roy turned? No way!*

Roy and his cohorts were caught by coincidence. They planned to pick the lock on the shoe store next door, then knock a hole in the wall, bypassing the perimeter alarm. From there they could simply climb through the hole into the jewelry store. Roy would then have the entire night to crack the safe.

Unbeknownst to the burglars, government agents were on the second floor of a building across the street, surveilling a business on the *other* side of the shoe store. When the agents saw what was happening at the shoe store and then at Cartier, they called the bobbies who swooped in to make the arrests. All five members of the gang received four-year sentences.[2]

Unlike Roy and Mike, I have never cracked a safe without consent. But my professional closet is not entirely free of skeletons. There is a small one from my first year of college.

I spent my freshman year at Western Washington University (WWU) with my lifelong friend Rick Wise. A wiry white kid with a big natural afro, Rick stood out in any crowd, and he walked like Denzel, probably before Denzel did. He was a blast to be around. Life for the two of us at WWU was one big party followed by another. And then came the midterm exams for which we were woefully unprepared. What to do?

To my eternal dismay, I rationalized unethical behavior and used my skills to cheat the system. The night before our first midterm, we waited until the buildings cleared out. When the coast was clear, I picked the lock to our professor's office door while Rick stood guard, ready to alert me if anyone ventured into the hallway. Once inside, we conducted a targeted search of the office and found not only the test but also its answer key. We copied it and marked the answers in code on folders that we laid at our feet during exams.

The heavy-duty locks at WWU used six-pin cylinders with multiple mushroom pins for pick resistance. I picked most of them in fifteen or twenty minutes, but a few were nightmares. One in particular, late in the spring, didn't yield to my magic touch until after three o'clock in the morning. And we didn't find an answer key. We found essay questions, which meant staying up all night to research our different answers — we couldn't turn in the same ones! We ended up prewriting essays in blue books that we smuggled into class the next day.

How many times did we do this? For every midterm and final the entire school year. There were occasional encounters with janitors and campus security, but we bluffed our way out of trouble by posing as teaching assistants. We eventually adopted the brazen tactic of leaving the office door *open*, ostensibly discussing some deep philosophical issue while "grading papers," oblivious to the outside world. We didn't look suspicious — we belonged there.

Our young and stupid selves did some young and stupid things. We picked locks and cheated on our exams. If the professors were grading on a curve, a few students might have received a slightly lower grade than they deserved. Hopefully, we didn't cause too much harm.

Some kids mature as teens and are ready for college. Clearly, I was not. And I was lucky not to have gotten caught — it would have altered the trajectory of my life.

Chapter Nine

Inverted Motherlodes

HEADING HOME FROM ALBERTO AND THE ATM, I REALIZE THAT I'M starving. It's time for a late lunch, so I drive right past my exit and head instead for downtown Camas. I pull into the drive-thru lane at my folks' Dairy Queen a few miles from the house with a serious craving for chicken strips. Mom makes them crazy good with a hot gravy dip on the side.

She greets me at the window with a big smile and a twinkle and passes me a cardboard container that makes me salivate like a Pavlovian dog. In her late seventies, my mother is still the most reliably positive and optimistic person I've ever known,[1] despite having had to endure every parent's worst nightmare, not once, but twice.[2]

"My dear son, would you like barbecue sauce or gravy?" This is Mom's playful way of reminding me that I resisted her gravy for years before finally trying it with chicken strips. It's delicious.

"If you're serving the gravy *cold*, ma'am, I'll take the barbecue sauce," I reply, winking and unable to suppress a grin.

Chuckling, she hands me a small container that's very warm, almost hot, and lays her hand on my arm. "Enjoy, David."

"Thanks, Mom. Love ya!"

"Love you more!"

With the gravy container wedged in my cup holder and a chicken strip between my teeth, I pull into traffic en route home. My phone rings through the speakers. It's an old acquaintance with Army Intelligence. He doesn't call often, so I chew quickly and swallow.

A weapons-maker colleague of his from southern Oregon has died suddenly, and two Graffunder gun safes need to be opened as soon as possible. The combinations were known only to the deceased. The family (or the government, it's unclear which) wants to charter a flight for tomorrow. A man from the Department of Defense (DoD) is en route to assist and oversee; my instructions are to pretend he is invisible. I can do that. Can I also be at Pearson Airpark at nine o'clock in the morning to meet my pilot?

I need to hurry this along before the gravy cools down, so I tell him yes, I'll be there, but at nine-thirty rather than nine. Val has morning appointments the rest of this week, so I'm taxi driver, and the extra thirty minutes will give me time to drop the kids at school.

Since I'm behind the wheel and can't take down any information, he's going to text me contact info for the pilot and the family, along with the gate code to get onto the estate. That text arrives as I'm devouring the remaining chicken strips and gravy. And a few fries. Well, *approximately* a few. Now I know where I'm going in the morning.

Opening a safe for a widow (or, less commonly, a widower) can be rewarding, but it's never fun. Amid great sadness, I have to ask for iden- tification and get answers to probing questions about executors, probate, and court orders. I sometimes feel like a nosy jerk, deeply invading the space of someone who has recently suffered a crushing loss. To mitigate the imposition, I try my best to be patient and respectful, to lend an ear when one is needed, and above all, to never be in a hurry.

Mr. Corey taught me long ago to avert my gaze. The best way to avoid being called as a witness in a court proceeding is to have nothing to offer. "If you don't look," he said, "you can't see. Capisce?"

That is good advice. Sometimes, though, the urge to peek is over- whelming. This is especially true when the client discovers an unexpected pot of gold and shrieks in delight, or, conversely, cries out in pain at uncovering bones that should have remained buried.

Mr. Corey and I once opened a safe for the family of the deceased mayor of a nearby town. The widow was in the house when we finally unlocked the safe in her garage. Her three children — two men and a woman, all in their late thirties or early forties — pulled open the door to reveal an extensive coin collection, assorted papers, a pistol, and a small recipe box. Lacking a container large enough to transport everything into the house, they laid it all out on a large rug in the middle of the garage.

The daughter, who appeared to be the oldest, knelt down and opened the box. It contained no recipes. She pulled out a handful of Polaroid photographs, took one look, and dropped them on the rug like hot potatoes. "Oh no," she cried out, covering her mouth with both hands.

"That can't be Dad," said one brother, staring down.

The other brother shook his head back and forth but said nothing.

Given the strong reactions, I couldn't help but steal a peek. Likewise for Mr. Corey. Each faded photo showed a different female performing a sex act on the same male. The women's faces were visible, but the man was guaranteed a certain level of anonymity since only a single anatomical appendage could be seen. Presumably, the widow would have had no trouble making positive identification.

When the elderly woman stepped out onto the garage floor and asked how things were going, one son scooped up the Polaroids, stuffed them back into the box, and walked over to Mr. Corey. The other son stood up and walked toward his mother to block her view. The daughter joined the distraction, talking animatedly about "Dad's coins."

The son with the box nonchalantly set it inside Mr. Corey's tool bag. "Will you *please* destroy these?" he asked in a quiet voice. Mr. Corey nodded and whispered, "Yes, of course."

There was absolutely no reason for the widow to see those photographs, and she never did. When we got back to the shop, Mr. Corey lit his butane torch, and we had us a stinky little bonfire.

Thirty years later, it was déjà vu all over again. I was contacted by John Davis, the owner of a very successful chain of locksmith shops here in the Pacific Northwest. He needed me for a trio of very heavy, difficult-to-open safes after the death of a prominent and wealthy businessman. It took me three days to open those monsters.

All the family members — and their lawyers — were present for the first safe, which was overflowing with paper money and coins. Jackpot. Most of them were still present the following day for the second, which was filled with guns and ammo. Only a few watchers remained on day three, and it was very late when we finally got the safe open. Inside were a few bank bags, a watch, some trinkets, and dozens of pornographic videos spanning a wide range of sexual proclivities. Judging by the covers, some of them likely included pedophilia and bestiality.

The watchers huddled. To my amazement, they asked us to take custody of the videos and dispose of them before any other family members showed up. While technically illegal, John and I agreed this was the right thing to do. There was no good reason for the elderly widow to see this stuff. And even if some of the videos were contraband, there was no perpetrator to prosecute — he was dead. We double-bagged that pile of garbage, and John incinerated it in the burn barrel on his property. Ashes to ashes.

CHAPTER TEN

The Bumble Bee Safe

A COUPLE OF MILES FROM THE HOUSE, I LISTEN TO A VOICEMAIL. IT'S FROM Rayna Apploff, the casting coordinator with Zodiac Media. She wants to discuss at greater length the possibility of a TV show on safecrackers.

We had a conversation about this awhile back, and I told her the same thing I said to Chelsea Stevens from TruTV, Doug Renfro from Radiant Media, Christie McConnell from Radley Studios, and Brooke Balick from Letter 10 Productions: I'm a little too old, a lot too introverted, and, frankly, I'm just not interested. Besides, most of my work is in financial institutions. There is absolutely no chance they would consent to video cameras rolling while I breach their security. I wouldn't even ask.

I'll email Rayna later to give her the names and numbers of friends in the industry who don't spend their days in banks and credit unions and are thus better suited for life on the small screen.

I've been on television a few times, mostly in short segments covering my safecracking classes around the country. These appearances are fun — except for the one that wasn't.

The first page of Google results for my name includes a news item about a nineteenth-century safe that supposedly "stumped" me. I'd like to pass it off as no big deal, but this on-screen episode caused me more than a little embarrassment. Here is what happened.

The long-shuttered Bumble Bee Cannery on historic Dock 39 in Astoria, Oregon, was being converted into condos when an old safe was discovered in the corner of an unused room. Floyd Holcomb and his associates at the Cannery Foundation wanted it open, so they made a few inquiries, and were advised to call me. Floyd asked if I could open it without drilling. I said the odds were reasonably good and quoted him a low price, provided he could wait until I was in the area on another job. Everything seemed fine.

The Bumble Bee Tuna safe at the Hanthorn Cannery in Astoria, Oregon.
Source: Author photo.

I showed up about a month later, after opening an ATM just a few blocks away. Floyd escorted me to an old Hall's cast-iron fire safe inside a large and long-defunct refrigeration unit. It was dark and cold, but he was kind enough to bring in a shop light and a portable heater. I unfolded my portable stool and sat down in front of the old safe, ready to play.

Many people were here before me, spinning the dial, so there was zero possibility that the safe was unlocked or that moving the dial a few numbers would do the trick. To open it without drilling meant either (a) going through the long list of tryout combinations and getting lucky or (b) manipulating the combination lock open.

When my hand touched the dial, I realized that my old book of try-out combinations was at home. I seldom use it because I can drill these antiques open faster than I can dial dozens or hundreds of possible combinations. But I found myself wishing that I had remembered to throw it in the van. The more options one has, the better.

With no tryouts to try, I'll either manipulate or drill. In most safes, manipulation involves looking for minute changes in the dial as various numbers are set. But on the particular type of lock found in this old Hall's safe, we use a different method. We look for tiny differences in how far the *handle* turns when the dial is set on various numbers. To make these differences easier to see, I attached a footlong pointer wire to the safe's handle. I then started dialing numbers and turning the handle, checking the tip of the pointer for any changes.

About an hour into the process, I heard feet shuffling on the floor behind me, followed by whispering among several people. In my work, this is a common occurrence. Passersby will often stop to watch the inaction for a few minutes before getting bored and continuing on with their day. I don't mind, so long as they don't distract or demand my attention.

I ignored the noise and kept manipulating until the room lit up like the Fourth of July. I swung around and was blinded by klieg lights pointing right at me. *This must be similar to the first few seconds of a rendition, I thought, right before you get manhandled, cuffed and bagged, and whisked away.*

An attractive, raven-haired woman walked through the wall of white. She introduced herself as a local news reporter and thrust a microphone in my face. *What the hell?* Then it dawned on me: Floyd had called the media. My savored solitude was over, unless I was willing to be a jerk and kick them out. Probably not a good idea, considering they were here by invitation.

No one had yelled "Action!" but the cameras were rolling, capturing my every move. Mr. Corey drilled into me, over and over: *never let the client see you sweat.* But in the heat of this particular moment, I found myself in the grip of an unfamiliar and quite uncomfortable feeling. I was rattled. Bigtime.

The reporter peppered me with questions, most of which ended up on the cutting room floor along with my inane answers. (Thank goodness for small favors.) It's still the closest I have been to an out-of-body experience. The word *surreal* is overused these days, but it accurately describes how this unfolding, dreamlike scenario felt. I was dizzy, slightly nauseated, severely cotton-mouthed, and really had to pee. It was awful.

Watching the late-night news that evening, I braced myself when the camera panned to me. Although I grimaced at the paunchy, balding, middle-aged safecracker who appeared on-screen, I was amazed that he came off as calm and reasonably articulate. He explained how the safe worked and what he was doing in general terms.

I watched as the guy on TV turned back to the safe and resumed manipulation. My hands were shaking, but you couldn't see it. I was sweating so profusely that it was running down my sides and back, but you couldn't see it. Miraculously, it didn't seep through my t-shirt and vest.

The hour I spent dialing with the lights, camera, and reporter in the room was a complete waste. There was no way I could dial that safe open in my condition. My fine motor skills were dulled, and my concentration was absolutely blown. I did my best — and I failed. It was time to implement Plan B.

I called Floyd over. Our entire exchange was captured on video, some of which played on the news. I calmly explained that, after two hours of unsuccessful manipulation, it was time to drill. Floyd, like many safe owners, resisted. "This safe is a historical artifact that should not be damaged with a drill," he said to the camera.

This was unsurprising. Safe owners frequently express concerns about drilling. They sometimes ask, wincingly, if it will ruin their safe. It absolutely will not, especially when the hole is tiny. I held up a few pediatric arthroscopes to the camera, along with a few 3/32″ and 1/8″ drill bits, to show how small the hole would be. I opened my repair bag and explained how easy it is to fill a hole with bearing balls and high-grade steel epoxy, thus making the safe far more drill-resistant than it was from the factory. Once the epoxy dries, I pointed out, the repair site can be made invisible with an acrylic paint pen and some smudge tricks.

I opened the database on my laptop to show Floyd and the camera a bunch of beautiful, same make and model antiques that had been opened with tiny holes. I zoomed in to show that the repair sites were undetectable.

I've had this exact discussion with thousands of safe owners. Once they understand the process, they want to move forward and have their safe opened — especially when the contents are unknown. It's human nature. With dreams of gold bars and fat stacks, people are always anxious to see what's inside.

Not Floyd. An eminently likable guy with a big mustache and an even bigger smile, he either was unimpressed with the presentation or had another agenda. He remained pleasant but shook his head back and forth while repeating the phrase "historical artifact" as a mantra. He wasn't budging.

I excused myself for a much-needed bathroom break, which gave me time to think. I needed a way to change Floyd's mind — I didn't want to leave with the safe still locked. Reentering the room, I called Floyd over for a quick discussion before the lights and camera could be turned back on.

I appealed to his sense of fairness. He had effectively ambushed me by inviting the reporter and her crew down to surprise me in the middle of the job. Letting me drill a hole the size of an uncooked spaghetti noodle was an easy way to even the score. He smiled and said "historical artifact" again.

I tried to gently dissuade him from by pointing out that this safe was hardly in pristine condition. The original flat black paint job, complete with gold leafing and mural, had long ago been painted over with Bumble Bee green. Over that was a thin coat of badly chipped Krylon glossy black. I knew it was risky pointing out flaws in his historical artifact, but I was running out of ideas.

Floyd crossed his arms, and I sensed he wasn't ready to deal. Before he could say you-know-what again, I decided to go all-in. "Tell you what, Floyd. Let me drill a tiny hole. It will be repaired better than the original. It will be invisible when I am done, and this will cost you *nothing*. My professional reputation is on the line, so this one is on the house."

Floyd uncrossed his arms and tapped his chin with his index finger. He appeared to be wavering. "How long will it take?" he asked.

I moved in for the kill. "Five minutes, mister. I can open this old Hall's in five minutes flat."

Big mistake.

"Five minutes?" Floyd asked, looking like he had just bitten into a lemon. I had a sinking feeling that I should have said thirty minutes or perhaps even an hour.

The lights and camera clicked back on as the reporter rushed over to our huddle. Before I could reply, Floyd recrossed his arms and shook his head. "I'm sorry, Dave, I can't do that. My board of directors would be upset if I okayed the drilling of this, um, historical artifact."

At this point, with the camera still rolling, I had two choices. I could continue dialing, or I could pack up my gear and call it a day. I gambled and chose the latter, hoping that Floyd would eventually get approval from his board and bring me back to drill the old safe.

Watching it on the news that evening was painful. Luckily, it aired at eleven o'clock rather than dinnertime. I went to sleep that night thinking there was a good possibility that no one saw it. I hoped that this embarrassing scene would recede from view long enough for me to get back to Astoria and finish the job.

Fat chance!

Woe was me, as my phone and inbox exploded over the next week. Articles about the safe and the "stumped" safecracker appeared in newspapers across the country. My parents called. My sisters called. My aunts and uncles and cousins called. Even my neighbors wandered over to inquire.

Everyone had the impression from the news stories that I had run into a safe I couldn't crack. Over and over, I patiently explained that this was an *easy* safe to open. I had pioneered a method on this exact make and model, requiring only a micro-penetration through the door. I had even written several journal articles on it. Everything would have gone smoothly, but the owner had bizarrely enforced a manipulation-only policy.

My father cut right to the chase. He asked if I had taken the job *knowing* that manipulation was the only acceptable method. I saw where this was leading, and agreed with Dad's implied point. If I want the flexibility to drill when dialing fails, I need the client to agree to my terms at the outset. Dad was right, of course, but it had never been an issue. I had always either succeeded at manipulation or persuaded the client into letting me drill.

Like every other profession, safecrackers have their chat rooms and discussion forums. My colleagues were aghast that such a lowly safe had "stumped" me. I was forced to endure a week or so of good-natured ribbing. I went along with it, but I'd be lying if I said it didn't irritate me.

A reporter from *The Oregonian* interviewed me, and he was surprised that I was surprised that Floyd hadn't called me back. He pointed out the prime-time publicity this had generated for Floyd, who was trying to sell the condos being built at the converted cannery. Aha! This explained why he had called the media in the first place, and why he was dragging it out so long. Over the next year, a parade of people traveled to Astoria to take a crack at that old Hall's safe. A host of locksmiths and safe technicians, a Navy safecracker, even a cryptographer from MIT made the trip to twirl the dial.

Finally, Tom Gorham, a locksmith friend of mine and club member whom I have helped on difficult openings in the past, succeeded where I failed — he dialed the safe open. Tip of the hat to Tom. In the big picture, it was a cheap lesson. The only injury was to my pride, which healed just fine in time, as nearly all things do.

WEDNESDAY

Chapter Eleven

Bank of America's Untutored Tenet

In the van, I tell the munchkins that I'm flying to a job after I drop them. Before they can protest, I promise to be back by suppertime. After all, on today's docket is a short plane ride and two gun safes. That's it.

They zero in on the keyword. "You *promise?*"

Hoist by my own petard, I grin and nod, and the kids file out of the van. I get three waves and three nearly identical half smiles as they skip-walk toward the entrance to the school. It was more rewarding when I got hugs, too, but this'll do. It's all part of the process.

Gotta hustle now. Val is making her famous chili and cornbread, and three of our children have basketball practice after dinner. Height-challenged *moi*, who wrestled in high school but never played b-ball, is the assistant coach and designated driver. I can't be late.

Heading out of the neighborhood, I listen to two voicemails. The first is from a law firm in Minneapolis. A very famous person has passed away. They won't tell me who it is just yet, but there is a bank vault in a mansion that must be opened soon, preferably Saturday. Well, that won't work. My daughter has a dance recital that I've already promised to attend. I call the firm and speak with an attorney who agrees to Sunday.

The second message is from Cortney Kinman, Diebold's regional manager for the western United States. A Cort call is a job call. Always. And this is an especially good one. Bank of America (BofA) is locked out of a Mosler bank vault in Albuquerque. They want to use this as a test case for potentially changing their policy.

Current policy requires that bank vault lockouts be solved by coring a man-size hole in the wall or ceiling. This permits someone to crawl through and open the vault door from the inside. Drilling the door — no matter how tiny the hole — is expressly prohibited. While I have drilled many BofA vaults around the country, each and every one has required layers of approval before finally getting the green light.

Sometimes, BofA has called me off the job *after* the go-ahead was given. In Kansas City, for example, they decided to cut a two-foot hole in the wall rather than have me drill the door. Problem was, the only access was an exterior wall. Everyone who drove by the bank that day observed a construction crew coring a hole through the outside wall of the bank.

One would think BofA wouldn't want to advertise quite so openly when their security is being breached. But that's not the worst of it. When it came time to repair the giant hole, they discovered that the custom stucco could not be matched and ended up replacing the entire wall. The total cost was more than ten times what I would have charged to drill and repair the door. Sad but true.

More recently, BofA put me on hold to gain yet another layer of approval after Diebold tech Larry Hedden picked me up at McCarran Airport en route to a vault drilling in Henderson, Nevada. This one put me in a bind since my return flight was in just a few hours.

Everything worked out. BofA got their act together and gave us the green light — Larry and I got the vault open in the nick of time, and I made it home that evening. But it put a few more gray hairs on my head. I'd really like BofA to change their policy so I can avoid this sort of high-pressure nonsense altogether.

There are many myths and misconceptions surrounding bank vaults. One of the more pervasive is that the doors are impenetrable and that coring a wall is therefore the best course of action. In fact, the opposite is true. No bank vault door is impenetrable, and coring a wall is almost never the best course of action. In nearly all cases, it's faster, cheaper, quieter, less messy, more secure, and far more professional to let an expert open the door. Let's look at these reasons in a little more detail.

First, speed and efficiency. The setup alone for a coring job takes longer than it does for me to drill the door. The crew will review building blueprints to determine the best place to breach the wall. They will bring in bulky equipment, run water lines for cooling the core drill, and lay down plastic sheets throughout the bank to prevent dust from damaging furniture as well as computers and other electronics.

Once setup is complete, they will begin the actual coring process, which can take all day or night. Most banks will close or have the job done after-hours due to the noise and mess. The repair takes even longer, which isn't surprising, given what is involved. They have to patch a man-size hole in a very thick concrete wall, reweld the rebar, replace the sheet-rock and paint, clean the carpet, and have a janitorial crew deep-clean the entire branch.

By contrast, my onsite setup time is about ten minutes, and I can usually open a vault door in two hours or less. Done and gone. This is made possible by decades of experience, the latest in high-tech equipment, and an extensive library of homemade schematics covering nearly all of the vault makes and models installed in the United States.

Second, ease of repair. There is an obvious difference between a quarter-inch (or smaller) hole in the vault door, and a twenty-inch (or larger) hole in the wall. To visualize the difference, consider this: more than five *thousand* quarter-inch bits can fit in a twenty-inch diameter hole. Common sense tells us that a humongous hole is much more difficult to repair. Common sense is correct.

Third, security. Drilling the door requires a properly equipped expert with detailed knowledge of make and model. This respects the layers of security in a meaningful way: When I drill a door, I'm breaching only *one* layer of security — I'm defeating either a combination lock or the timelock. *But not both.* Even if I were to leave my tiny hole open, no one, not even another expert with medical-grade scopes could use that hole to open the vault. Coring a wall bypasses *all* the layers of physical security. And once a wall is cored, its integrity is forever compromised.

Fourth, and finally, the cost. A core drilling job and the associated repair cost many thousands of dollars. Moreover, there is an element of risk in coring a wall or ceiling. I've seen damage to alarm and electrical systems, building utilities, safe deposit boxes, flooring, and more. A coring job gone wrong can turn into a nightmare, and when it does, the cost and hassle quickly spiral upward. A qualified and competent safecracker costs far less, and our methods pose no risk of damage to the building's infrastructure.

Given the slam-dunk case for drilling the door, you might wonder why BofA prohibits it. The answer is not *totally* hare-brained. Their anti-drilling policy is rooted in a legitimate concern for aesthetics, but also in an illegitimate and outdated understanding of how the vault rating system works.

I share BofA's desire to avoid ugly battle scars on a vault door, and I employ many tricks to achieve an aesthetically pleasing result. Some doors have a thin, stainless steel skin that can be detached before drilling. I remove the skin, drill the hole to open the vault, patch it better than original, and reinstall the skin. On other doors, I remove a hinge cover or handle escutcheon or dial ring, and drill underneath. Sometimes I insert a robotic arm through the mandatory ventilation hole and open the vault door without drilling any holes at all. The goal is to leave no visible evidence that the door was penetrated. Always. And where there's a will, there's a way.

BofA's other concern is that if a vault door is drilled, its rating from Underwriters Laboratories (UL) will be lost.[1] Their concern here is misguided; they simply haven't kept up with developments at UL.

In brief: UL tests safes, vaults, and vault panels for burglary resistance. The higher the resistance, the higher the rating. These ratings are translated into what UL calls "Listing Marks" that are inked or etched onto labels that are attached to the products before they leave the manufacturing facility.

A safe sporting a TL-30 label, for example, offers more burglary resistance than one with a TL-15 label. Likewise, a vault with a Class 3 label provides more protection than one with a Class 2 or Class 1 label. Financial institutions, jewelry stores, insurance companies, and other informed consumers rely on UL's rating system as a guide for purchasing and coverage.

At one time, there was confusion over whether drilling a security container voided its UL Listing, and, if it did, whether a service technician was obligated to remove the label.[2] But UL long ago posted a clarification on their website and put the issue to rest.[3] The upshot is that UL is utterly agnostic about whether a drilled and repaired container continues to provide the same level of burglary resistance it did the day it rolled off the assembly line. This is common sense — the folks at UL can't know what they don't know.

Apparently, the decision makers at BofA are unaware that they are contradicting UL on UL's policy. Moreover, they don't appear to grasp the implications of their own position. For example, many vault doors currently in banks across the United States aren't UL-Listed, so *there's no Listing to be voided.*[4]

Of the remaining doors that *are* UL-Listed, many of them are used on vault rooms whose walls are made from panels that are also UL-Listed. The logic here is straightforward. If a quarter-inch hole in a Class 1 vault door voids its Listing, then a two-foot hole in a Class 1 vault panel (the wall) must void its identical Listing.

In truth, neither voids anything. It comes down to the quality of the repair. A competent professional makes a clean penetration, effects a better-than-original repair, and strives to make it invisible. For an expert, this is all in a day's work.

After merging onto the freeway, I call Cort and get right down to business. "Hey mister, how soon do you need me in Albuquerque?"

"Well, BofA wants to make sure the timelock isn't overwound before they spend the big bucks to fly you in. They've been locked out all week, Dave, and they'll know for sure Friday morning. I'll call you either way. If it doesn't open, you can catch a flight early next week."

"That's perfect, Cort. I'll hold Monday open."

I'm excited. *I want this job.* I'd love to bring BofA over from the dark side and show them all the advantages of having me drill the door.

Fingers crossed.

CHAPTER TWELVE

A Chartered Flight

WITH CRUISE CONTROL AT SIXTY, I'M READY FOR A PODCAST. SOCIALLY liberal and fiscally conservative, I like to hear what *both* sides are saying. So, what's it gonna be: conservative Ben Shapiro or liberal Pod Save America? I enjoy them both, but my ringing phone says neither.

It's Jim Wicken from Cook Security Group, a regional bank service competitor to Diebold. Nearing retirement, Jim has been in the business even longer than I have, and we've worked together many times in the Pacific Northwest.

Today it's Wells Fargo with a problem. They are locked out of a Diebold Guardian bank vault at a branch within driving distance of my home. The Guardian utilizes one of the slickest designs in safe and vault history. Like nearly all bank vaults, it uses two combination locks. But unlike almost any other, it has only one dial. Through a triumph of engineering, two completely independent locks are controlled by a single dial.[1]

Jim wants me to open it *today*. When it rains, it pours. I explain my schedule and urge him to push for tomorrow, which is wide open at this point. It's rare for a bank to insist on a same-day opening. It already happened once this week; there is no way it will happen again.

He will make a call and get back to me. Tomorrow is going to be challenging but profitable. Diebold's Guardian is number one on my list of difficult-to-penetrate vaults, and I charge accordingly.

Pulling into the lot at Pearson Airpark, I spot the fence and head toward the gate. My pilot, Monte, waves me onto the tarmac, and I park alongside the corporate Cessna. I don't know if it was the family or the government that chartered this flight, but I'm grateful either way.

Jim calls back as I finish loading tools. Wells Fargo cannot wait. A high-profile client needs to access a safe deposit box inside the vault and snag her passport for an international flight departing at dawn tomorrow. It's today or bust.

Unbelievable.

It's bust, then. A coring crew will bore a gigantic hole in the wall, and Jim will crawl through and open the vault from the inside. It pains me greatly when this happens, but my hands are tied. Today I'm committed elsewhere.

It sucks to lose a good job, but there is an upside. Chartered flights have no security lines, no waiting, and no restrictions. I cue up a mellow Eva Cassidy playlist and raise my laptop screen while we taxi down the runway. In five minutes, Monte and I are in the air above Portland, heading south. Man, I could get used to this.

I'm banging away on the keys when Eva begins her fingerpicking introduction to one of my favorites. I turn the volume up a little, close my eyes, and take a short break.

In Simon and Garfunkel's hands, "Kathy's Song" is a great tune. In Eva's, it's hauntingly, achingly beautiful. A masterpiece of simplicity, it contains no other instruments. It's just Eva and her guitar, recorded live by a friend who brought a tape recorder to the little restaurant where she occasionally played for a few bucks and a burger.

Like many people who have been deeply moved by Eva's music, I remember the first time I heard her voice. I was writing at the dining room table when Val walked quietly into the kitchen, smiled, and put a CD in the stereo.

I rarely notice low-volume, laid-back music when I'm working. I can usually tune out the world. Not this time. Thirty seconds into a cover of Sting's "Fields of Gold," I stopped and tilted my head toward the source. This voice and delivery demanded my immediate attention.

"Wow," I said, turning in my seat. "Who *is* that?"

Val nodded. "I knew it!" she exclaimed. "I *knew* you would react the way I did. Her name is Eva Cassidy."

Nearly all her recordings were covers, but Eva had a way of climbing inside the soul of a song and making it her own, causing many listeners to forget all about the original. From Judy Garland's "Over the Rainbow" to John Lennon's "Imagine", Sarah Vaughn's "You've Changed" to Fleetwood Mac's "Songbird", Eva did it over and over again. What a gift.

We hit some turbulence, so I back off the keyboard for a minute. I'm writing on a pair of complicated doors: the new AMVault from American Security Products, and a first-generation panic-room vault door from Graffunder. Both gave me fits earlier this month, and I think my fellow safecrackers will appreciate the heads-up on a couple of difficult openings.

All high-end safes and vaults are equipped with safeguards that protect against a variety of attacks. One such safeguard is the *relocker* — a device that prevents the door from being unlocked (or pulled open) after it's damaged in certain areas. To avoid firing relockers, I'm scrupulously careful where I drill.

The redesigned AMVault caught me by surprise. I drilled through the door and defeated the lock only to discover that I had inadvertently fired not one, not two, but *three* relocking pins that prevented the boltwork from retracting. It was ugly.

I wrote a tongue-in-cheek note that evening to the lead engineer at American Security Products, whom I have known for more than two decades, thanking him for (not) keeping me in the loop. I would rather learn about design changes *before* I have to deal with them.

A recent encounter with a Graffunder vault door guarding a panic room proved to be another mind-bending challenge. Built on the same chassis as Graffunder's elite-level gun safe, this door incorporates high-quality steel, a premium hardplate, and the industry's best relocking system — during a burglary attack, it prevents the handle from turning *and* stops the door from swinging open.

The in-swinging vault door must be open before the panic room can be accessed. Enter the room, push the door closed, turn the interior handle to extend the boltwork, and then turn the key to lock the boltwork in the extended position.

If you fail to turn the key from inside the panic room, the bad guys chasing you can simply rotate the vault's handle to gain access and ruin your day. It's a good system, so long as you remember to lock yourself in.

My client inadvertently bumped the key as he was exiting his panic room. Once the door was shut and locked, it wouldn't open.

The combination lock worked fine, but the keylock had partially locked when the key was bumped. Since the keylock was *inside* the locked vault, there was no easy fix available. It was my job to figure out where that keylock was located and devise a way to defeat it. With no external keyhole, this was no easy task. I spent half an hour scaling from photos of a similar panic-room vault, figuring out approximately where to drill, before I chucked up a bit. The doodling time paid off. Two small-diameter, easily fixed holes were all it took to open that monster. Definitely worth a long article.

We touch down in Medford, and, while the Cessna's wheels are still rolling, I read an email from my oldest son, Justin. A few years ago, I pitched Diebold a plan to train their entire workforce in basic safe-cracking and provide them with online technical support. Justin, an accomplished computer programmer, built a beautiful forum with drag-and-drop photo uploading and a host of other user-friendly features. I demonstrated it live during a presentation at Diebold's headquarters in Canton, Ohio.

Diebold liked the concept but not the price, and the forum has been sitting dormant in a remote corner of the web ever since. Justin asks if I've considered "rebooting and rebranding it," and, if so, "what's a good name?"

Wow, I love it — revive the forum and provide tech support to locksmiths and safe technicians worldwide. It might provide a gradual path to semiretirement. I could never *fully* retire from the field, because opening safes and vaults is too much fun, but I'd welcome a reduced workload.

What to call it? Obviously it can't remain the Diebold Tech-Support Forum. Hmmm, maybe The Safecracker Support Forum?

Has a nice ring to it, I think.

CHAPTER THIRTEEN

At the Estate: Two Tough Safes

I TAKE A DEEP BREATH BEFORE PUNCHING IN THE GATE CODE AND HEADING to the primary residence. A death in the family usually elicits one of two responses, each revealing the character of the surviving relatives. Some heirs suffer paralyzing grief. They often alternate between humorous anecdotes and great sadness. Others are giddy with greed, concerned only with the goodies that will be theirs once the safe is open. Money seems to bring out the worst in some people.

To my relief, today's family is the former. None of them has any idea what's inside, but they are utterly uninterested in the contents beyond the legal necessity of taking inventory. In the background is a man from the DoD. He doesn't identify himself, and I don't ask — not knowing certain things is part of my job. I can only assume it has something to do with the weapons produced by the deceased's company.

In safes and vaults, size is almost irrelevant to security. The smallest safes can sometimes be the toughest to crack. Today, the upstairs safe is a tough and thick Graffunder, about five feet tall. Downstairs is a tougher and thicker Graffunder, about six feet tall.

I'm often asked by friends and acquaintances for a gun safe recommendation. For a long time, professional safecrackers considered gun safes to be a joke. Most were built by relatively new companies with little experience in the industry, and they contained weaknesses that could be exploited by children with access to a few tools or a little luck. Unfortunately, many mid-level and nearly all low-end gun safes still exhibit those same vulnerabilities today. But not Graffunder.

This Graffunder gun safe houses my tools each night after work. It gets an annual checkup because a lockout is unacceptable — it would require me to open the safe to get my tools to open the safe.
Source: Author photo.

A simple rule of thumb: if a gun safe is on the floor of a big box retailer for a few hundred dollars, it's a toy. Don't waste your money. Despite the grandiose claims made on their labels, such safes offer very little protection from burglary and even less from fire. For concrete proof, go to YouTube and watch a few videos of junky gun safes being forced open in less time than it takes to read this page. Simply put, decent security is not cheap.

Neither of today's safes is cheap, and both provide excellent security from fire *and* burglary. They are similar in appearance, except for color and size. I decide to attack the shorter one first.

I expect a bearing-ball hardplate, which is composed of chrome steel bearing balls sandwiched between two plates. The plates can be either aluminum or steel, and the better ones are shaped like egg crates that leave each ball isolated in its own compartment, but loose enough to spin, snag, and snap off trespassing drill bits. The bane of novice safecrackers everywhere, bearing-ball hardplates are now used in the majority of burglary-resistant containers.

Before tangling with bearing balls, though, I want to give manipulation a shot. Manipulation is the scientific art — or artistic science — of using clues from the dial to determine the combination. Movies usually portray this through a series of clicks as the dial is turned, followed by a louder *clunk* when the lock opens. It looks and sounds good on-screen, but it's a total fabrication.

On most combination locks, we look, listen, and feel for little bumps called *contact points* as the dial is rotated. Sometimes contact points emit a detectable sound, sometimes they don't. It all depends on the particular lock and the surrounding environment. We note the dial number where contact occurs down to one-tenth of a digit, and watch for changes in that fractional number as we set the tumblers on different combinations.[1] Those changes are the key to success.

I manipulated weekly when mechanical locks dominated the market, but it's once or twice a month these days. It works *only* on safes with a functioning mechanical lock whose combination is unknown. It doesn't work on the electronic locks that are incrementally taking over the market, or on malfunctioning safes, and it's often too time-consuming to be practical on most manipulation-resistant locks.

In the right environment, manipulation can be a joy. But today's opponents are in dark locations with glossy, high-glare dials and very spongy contact points. Not fun.

The safe downstairs is a slightly better candidate, so I tape one of my LED light sources above the dial. Then I place a no-glare plastic strip over the opening index on the dial ring, and a matching one on the dial. This helps a little, but the glare off the dial and dial ring is still excruciating, even with a tissue over the LED to diffuse the light.

I know from experience that this will be a tedious chore and a strain on the eyes. Bag it, then. I'd rather drill.

I step back from the shorter Graffunder and give it the once-over. This is an attractive safe: glossy black metallic finish with matching chrome dial and "L" handle. Silver on black, the same color scheme I chose for the Graffunder that sits in my garage and houses my safecracking tools each night after work.

I have multiple options on most safes: I can penetrate the top or side, and then use a long bronchoscope to peek inside the lock. Or I can drill through the door directly into the lock, and then use a short arthroscope to view the tumblers.

Both of today's safes are recessed into beautiful, custom-built cabinets, so I can't drill through the top or side without first pulling out the safes. I dismiss this idea immediately. I learned long ago that safe *cracking* and safe *moving* require two very different skillsets, one that I possess and one that I absolutely do not.

My Two Best Attempts at a Darwin Award

When I was working for Mr. Corey, he bought a gigantic Mosler double-door antique banker's safe with three interior jeweler's chests. He estimated its weight at more than six thousand pounds.

When the delivery truck arrived, we went out to inspect the safe and make sure it hadn't been damaged in transit. Satisfied, Mr. Corey and the loquacious truck driver used a couple of Johnson Bars to push-roll this statuesque stunner (on casters) onto the heavy-duty liftgate.

I hopped down, while Mr. Corey stayed up top with the trucker who hit the lever to lower the liftgate. Chatting, neither of them noticed: although all four casters were on the liftgate, the safe's lower edge wasn't quite clearing the bed of the truck. While the liftgate and one side of the safe were going down, the other side was riding on the edge of the bed, causing the mammoth Mosler to tilt forward, toward me.

I stood motionless as the safe tilted more and more. I finally reached out to push back, crazily thinking I could stop it. When Mr. Corey noticed, he barked loudly, "STOP STOP STOP!"

The truck driver released the lever, but it was too late — the safe was in a slow-motion fall, while I was doing the limbo, trying to prevent three tons of steel from following the laws of physics. I finally fell backward, hit the ground, and rolled quickly to my right. Being an in-shape high school wrestler probably saved my life. That old Mosler fell through the sidewalk right next to me, breaking the concrete out in a jagged mold around it.

"Dave! Dave!" Mr. Corey shouted, unable to see me. I was momentarily in shock, pinned against the safe. When I tried to back away, I couldn't move. The safe had come down on the corner of my letterman's jacket, and when the concrete broke away, the jacket went down with the safe, pulling me hard against it.

I rolled the other way, out of my sleeves, and scrambled to my feet. When Mr. Corey saw my head rise above the cloud of dust, he leapt off the side of the truck and bear-hugged me like a long-lost son. The jacket was ruined; I was embarrassed but unharmed.

Fifteen years later, I made another run at a Darwin, and this time I didn't escape unscathed. It was going to be a busy day. I had five openings scheduled: one in Salem, two at the same Portland location, one in Gresham, and one in Kelso, Washington.

Salem and Portland went bam-bam-bam. Had them all open before noon and headed to Gresham. I walked into the warehouse and discovered the safe, laying on its side, facing the wall. It needed to be stood up and turned around for me to work on it. But the forklift driver was at lunch.

I should have left and come back another day. Instead, I grabbed a Johnson Bar and leveraged that eight-hundred-pound heavy safe up to the tipping point. Then, for some unknown reason, I slipped and fell. So did the safe, which hit the ground beside me.

I stood up to dust myself off. *Whew — that was close!* And then I saw blood. What the heck? Where was it coming from? My legs were okay. My arms were attached. My left hand was intact. But the ring finger on my right hand was nearly half gone. What remained was a bloody stump.

I hadn't felt a thing. The hand surgeon later told me that I was lucky for two reasons. The corner of the safe had severed only part of one finger, and the nerve had come out with the partial finger that remained under the safe. Yes, I had to wedge the safe up a few inches to retrieve that flattened piece of hamburger for the trip to the emergency room.

Two surgeries and three months later, I was as good as new. I went back to the warehouse and got my revenge on the safe that took my finger, and later wrote a fess-up article on the whole episode. My publisher titled it "An (Un)Safe Opening Job."

Unsafe indeed. I can be a little slow on the uptake, but I eventually learn the essential lessons. This one taught me not to climb back on the safe-moving horse. And I haven't.

Because of my failures as a rigger, the decision not to move the safes out of their cabinets is a no-brainer. Graffunder puts the kibosh on top and side drilling anyway. They surround the lock with steel plates and use a large Allen set screw to block visual access through the back. This is high-level protection that very few gun safes on the market can match.

I have to go through the door and its hardplate. Two methods to choose from. I can remove the dial and drill straight in, or leave the dial alone and angle-drill through the dial ring. Since there is a decent chance that both safes are set on the same combination, I decide to drill through the dial ring and leave the dial intact. This will simplify the process of deciphering the existing combo.

With Plan A set, I open the Beemer case, pull out a large shop rag, and tape it to the floor in front of the safe. I plug in the Fein motor, chuck up a 5/32″ bit, and commence drilling. I'm penetrating low on the dial ring and aiming up at a fairly steep angle. Penetration proceeds smoothly until I hit the hardplate. I peek in with a pediatric arthroscope and spy the edge of a bearing ball along the extreme right side of the hole. I drop down to a 1/8″ carbide-tipped specialty bit and gently nudge it to the left. It grazes the ball on its way past. Home free.

The Fein lurches forward slightly as the bit enters the lock. In the hole goes the scope, and oh, what a beautiful sight: I can see all three tumblers. I turn the dial carefully and write down the number at which each tumbler is aligned correctly. Once the lock-bolt retracts, I rotate the safe's handle and the first Graffunder is unlocked.

"It's ready," I announce, taking a step back. Carrying a thick briefcase, the man from the government walks up and pulls open the door, takes one look inside, and backs away empty-handed. Whatever he's after isn't in this safe. He nods to the family members as I head downstairs to try the combination on Graffunder number two.

No luck. I reverse the combo — a common trick — but that doesn't work either. I try all six possible permutations of the three numbers, checking the contact points each time. Nothing. Graffunder number two is set on a completely different combo. *Rats.*

They're done with the upstairs safe, so I plod back up to repair the drilled hole, replace the dial ring, and set a new combination. Most people don't like complicated combos, so I'm always careful to select one that's easy to dial. This means setting two of the three numbers on 0s and 5s. Or, in the case an elderly person who has trouble seeing, all three. What makes such combos easier are the long hash marks on every 0 and 5; they are much easier to see than the smaller, numberless hash marks that represent every other digit all the way around the dial.

Some examples of easy-to-dial combos are 20–80–35, 55–10–44, and 66–5–30. At least two of the three numbers are 0s or 5s. They aren't so close together as to cause users to dial past a number, and not so far apart as to confuse users on how many rotations they have dialed. Like baby bear's porridge, these combos are just right.

The potential downside is that, by limiting combos to those that are user-friendly, we decrease the possibilities from a theoretical one million to fewer than fifty thousand. While this is a significant reduction, the real-world risk of a burglar dialing tens of thousands of combos is likely very close to zero.

Heading back down the stairs, I glance at my phone. Five missed calls, two messages, and goodness, it's nearly one o'clock. What the hey?! I've been lulled into a work trance on many occasions, only to wake up and discover a big chunk of the day has disappeared. Just a Rip Van Wrinkle in time, I s'pose.

Graffunder number two is a beauty: metallic forest green door and body with complementary brass trim. If safes were models, this one would be walking the runway, posing for coveted cover shots, and vacationing in the Hamptons.

I had good luck upstairs with angle drilling, so I opt for the same attack here. I slip a business card in the jamb and discover that the door slab on Graffunder number two is exactly twice as thick as the slab on Graffunder number one. This means I can lessen the drilling angle to hit the same target. This is good. The lesser the angle, the lower the risk of missing the mark or damaging a component.

Once again, a large shop rag is taped to the floor, and a 5/32″ bit is chucked up in my Fein. I choose a spot low on the dial ring and commence drilling. I do my best to place this hole a tiny bit further left than before, hoping to thread the needle and go between the bearing balls. But I hit the edge of a bearing ball along the extreme right side, just like last time.

I again drop down to a 1/8″ carbide-tipped specialty bit and gently nudge it to the left. It just grazes the chrome steel bearing ball on its way past. But instead of the bit continuing on and drilling into the lock, it stops cutting. Even more surprising, a puff of gray smoke rolls out the hole.

This is wrong. I pull out the bit to examine the tip, but all that's left is a hot bluish blob. *Huh?* Bearing balls don't melt carbide tips turning at low speed. They snag and snap them.

I ease a new 1/8″ bit in until it hits the same spot and stops. I remove the bit from the hole and more smoke rolls out. A look at the tip reveals another melted blue blob. Enough already!

Time to go bigger. A brand-new 3/16″ enlarges and cleans out the hole nicely. Looking in, I see a second piece of hardplate, stacked right behind the first, and it is super hard. *Great.* I have a decision to make: continue on with this angled hole through the dial ring or cut my losses and drill a new hole straight in with a bigger bit.

Safecrackers are reluctant to give up on a hole. Drilling a second one feels like failure, but sometimes it's the most prudent course of action.

This is one of those times, because staying the course will take too long and require more bits than I care to expend. In addition is the aggravation factor, looming larger with each passing year, and often triggered by things like melted blue blobs.

I move to Plan B and grab my specialty tool for removing dials. Dial removal can be tricky. You have to know the correct method for each of the many dials in existence. The wrong method can break a critical component in the lock and needlessly complicate the opening.

The best all-around dial puller is made by a friend of mine in Australia. He's a retired safecracker who makes quality tools and sells them to working pros to pay his green fees on golf courses around the world. This bloke is doing retirement right! His dial puller works like a wheel puller or gear puller — a jackscrew down the center separates dial from spindle with no damage to any components.

It only takes a minute to remove the dial and ring and get started. I carefully measure out the drill point and get to work. Thirty minutes later, I have a hole through both hardplates and into the lock. My arthroscope goes down the hole and the view is scrumpdillyicious. I align all three tumblers, turn the spindle until it stops, and rotate the handle to unlock Graffunder number two.

"It's ready," I announce, stepping back and busying myself with the cleanup.

The man from the government walks over and pulls open the safe door. I'm trained not to look, but he's making no attempt to shield what he's doing. I accidentally glance over just as he removes a short stack of Manilla folders from the top shelf, places them in his briefcase, and clicks it shut. He turns in my direction and nods as he walks past.

I patch the drilled holes with multiple bearing balls, pieces of broken taps, and premium-grade steel epoxy. Anyone who tries to penetrate my repair is in for a most unpleasant surprise.

My replacement dial is flat black with white numbers. Not a match to the original gold, but its antireflective properties cause less eye strain, and the numbers are much easier to dial. I set a user-friendly combination, collect payment for services rendered, and say goodbye to a good family and the enigma carrying the briefcase. He smiles mutely in return.

I'm curious, but I'll never know who he was or why he was there. Some mysteries must remain mysteries.

Monte is happy to see me. He wants to get airborne as much as I do. I stow the Beemer case, grab my laptop, and in minutes we are high above southern Oregon, heading north for the airpark and home.

I busy myself with writing. First drafts of both articles are completed by the time we land in Vancouver. For me, this kind of output is nearly unheard of.

I need my own plane and pilot!

Chapter Fourteen

Saying Goodbye to Mr. Corey

AFTER DINNER, VALERIE AND I CLEAN THE KITCHEN WHILE THE BOYS get ready for practice. She and I met on the debate team at itty-bitty Clark College, where we were national champions, taking home gold in team debate, and gold and silver in Lincoln-Douglas. Val went on to do the same thing in law school, winning the U.S. championship in Moot Court Nationals with her two NYU teammates, walking away with the awards for Best Appellate Advocate and Best Oral Advocate all by her lonesome. She is fun and formidable in discussion and debate, and then some.

The boys and I walk out of the house toward the van carrying basketballs, orange cones, and a clipboard. Val waves goodbye and closes the door behind us.

I was head coach for two seasons, back when the focus was on fundamentals — dribbling, passing, and layups. But when it was time to work on screens and complicated plays, I passed the baton to Alex, who plays on the high school team and whose knowledge of the game far surpasses my own. I basically follow his lead and help maintain order. Assistant coach fits me perfectly now.

Alex twists in the passenger seat to face his younger brothers. "You guys are gonna demonstrate the give and go tonight, okay?"

"Okay," says Aidan. "I'll be the giver and goer."

"No fair," says Julian. "It's my turn!"

"How about you guys trade-off," Alex interjects. "Waddya say?" They both shrug-nod.

My phone is sitting in its holder when a notification lights up the screen. Alex notices and tilts it toward me. "Hey Dad, who's Pastor Rick?"

"Rick Warren is a Christian leader who wrote one of the most successful books of all time."

"Have you read it?"

"Sure have. It's on the religion shelf in my office if you want to browse it."

I sometimes listen to Pastor Rick's podcasts. He's the founder of Saddleback Church and author of the mega bestseller, *The Purpose-Driven Life*. Although I don't share his core beliefs, I like his relaxed presentation style. His demeanor is in stark contrast to the televangelists and megachurch pastors whose fleecing histrionics grate on me like fingernails on a chalkboard.

Warren weaves personal stories into his teachings so well that the listener soon feels they know him. He leaves the impression that there isn't much difference between the private and public versions of Pastor Rick, that what you see truly is what you get. He's the kind, wise neighbor, chatting with you over the picket fence.

The first time I encountered Warren, I was dumbstruck by how much he reminded me of Mr. Corey — their voices and speaking styles were so eerily similar that I don't know if I could have distinguished them. Confident and charismatic, but in a low-key way.

Strong, gentle souls both.

When Mr. Corey was diagnosed with cancer, he and Teddy moved back to his native Spokane. I knew he was terminal; he knew my first book was about to be published. So when he asked if I would bring him a signed copy hot off the presses, I said yes without hesitation.

"Promise?" he said, softly into the phone.

"Yes sir, absolutely."

I begged my publisher to expedite printing and overnight me a few copies at whatever cost. FedEx arrived a few weeks later around ten o'clock in the morning, and I was on the road to Spokane five minutes later.

Mr. Corey was the first adult in my life who insisted on being called by their first name — even by a school kid who'd been taught to address adults as mister and missus, or sir and ma'am, and couldn't shake the habit.

"Dave," he would say, drawing out my name in mock exasperation, tilting up the Optivisor that always circled his head at the workbench.

"Yes sir," I would reply, looking over, knowing full well what was coming.

"*Mister* Corey was my *dad*. For the *nth* time, son, please call me Gene."

"I'll try, sir."

Well, I did try. But it never felt right. I continually relapsed, and he eventually relented. Deep down, I think he liked being called mister.

Mr. Corey's wife, a lovely and petite Japanese lady with beautiful salt and pepper hair that fell to her waist, was the second adult in my life who insisted on being called by their first name. The discomfort she displayed at being called missus or ma'am made it clear that she was serious.

Problem was, her given name was Teruko, which none of us could pronounce. That clicky "R" sound is difficult for us native English speakers. Everybody at the shop, including Mr. Corey, called her Teddy instead.

I worked after school on weekdays (in the wrestling off-season) and all day every Saturday. At lunchtime on weekends, Teddy made tuna fish sandwiches on brown bread with diced onions, a little sweet relish, and a layer of crunchy lettuce, finished off with a light dusting of parmesan. My mouth would water just watching her make them. Teddy's Tunas were to die for.[1]

I hit the outskirts of Spokane in the late afternoon, gassed up the van, and arrived at the Corey home a few minutes later, a little nervous, not knowing what to expect. I hadn't seen Mr. Corey or Teddy in more than five years.

Teddy met me at the front door and urged me to prepare myself, for Mr. Corey no longer looked like the man I once knew. When I worked for him, he was a robust two hundred and forty pounds of high-energy locksmith with a ruddy complexion and a positive, can-do attitude that never flagged.

When I walked into the bedroom, he was propped up in a hospital bed he wouldn't leave alive, with wires and tubes everywhere and skin the color of split-pea soup. I was overwhelmed with grief on the inside, but on the outside I kept it contained, just barely.

He had lost more than a hundred pounds while the cancer ravaged every corner of his body. And still he smiled. He went ear to ear when I laid a signed copy of *The National Locksmith Guide to Safe Opening* on the large tray straddling his lap. He sighed and slowly turned its pages. "This is nice, son," he said in a whispery rasp. "Really, really nice."

"Thank you, sir," I said, trying to wish away the welling in the corner of my eyes. "Well, I'll be," he said, pausing at photographs of an antique safe toward the front of the book. "Isn't this the one that kicked our butt and then miraculously popped open?"

"Yes, it is," I replied with a smile, relieved to be reminiscing rather than facing reality.

We had worked on that old Diebold all day long, but couldn't figure out why it refused to open. Mr. Corey finally insisted that we break for an early dinner. A pause in the action often improves morale and yields fresh eyes.

I had been laying on the floor for hours, fiddling and probing and peering down the hole we drilled. Food sounded good, a break even better. My left leg was a little numb, so I reached up and grabbed the safe's handle to pull myself up. To our shock, the handle turned — the safe was open!

The client was watching, so Mr. Corey simply said, "Good job, son," as if that had been our plan all along. He was a great role model for keeping one's cool, not only when things go south, but also when they veer unexpectedly north.

We reminisced about other amazing and lucky jobs, about the botch-jobs we would both like to forget, and about my near-death experience on the sidewalk in front of the shop. He just shook his head. "I thought you were a goner, son."

Teddy came over and asked if I had eaten dinner. I hadn't. "What can I make you?" she asked.

I looked over at Mr. Corey, then back at Teddy. "Do you remember Sandwich Saturday?"

Mr. Corey howled hoarsely. Teddy, visibly pleased, smiled widely and said, "Of course, David." And off she went to the kitchen.

She returned shortly, hoisting a serving tray on which rested a single Teddy's Tuna and a tall glass of iced tea. Mr. Corey anticipated my concern and whispered, "It's okay son, I can't eat solid foods anymore. But you go right ahead. Please."

The first bite was taken in an uncomfortable fishbowl, but the sandwich tasted exactly as I remembered. "Teddy," I said, "your tuna is still the champ." She tilted her head toward me and closed her eyes for a second in a silent thank you. Mr. Corey continued flipping through the book, smiling as a few more memories appeared on the pages before him.

Teddy took the plate away when I finished eating. I gulped down the rest of the iced tea, steeled myself, turned to face the bed and said, "Mr. Corey, I need you to know something."

Having long ago given up any hope of ever getting me to call him Gene, Mr. Corey just looked up from the book, and I continued my short, memorized speech. "Without your encouragement, especially given how mechanically inept I was when we first met, this book would *never* have been written. I wouldn't have made it in an industry that places such a high value on traits I didn't naturally possess. You were a great mentor. Not only on job tasks, but also on how to treat people and how to present yourself in public."

He slowly raised his hand to stop me. "David," he whispered hoarsely, shaking his head. "You started slowly, yes, but once you began figuring things out, especially when it came to safes and vaults, you went from zero to sixty in a blink."

He rested for a moment without looking away. "That was you, son," he said about as sternly as he could, given his condition. "Capisce?"

I nodded reflexively, but Mr. Corey was wrong. I wrestled mentally with how to explain it. Without his patient prodding, I would have crashed and burned in my very first year and left the trade. He instilled in me a never-give-up attitude. He pushed me to overcome my mechanical deficit by taking photographs and documenting every single safe I could get my hands on, working out precise drill points for as many different makes and models as possible. This required leaving his employ to work for a large safe and vault company, which I eventually did, begrudgingly.

I was about to object gently when he whispered, "Teddy?" He pointed to an old roll-top desk against the wall. She nodded knowingly, walked

84

over and opened the upper-right drawer, and removed a small looseleaf book with a weathered, plain black cover. I recognized it immediately.

Mr. Corey's father started Corey Safe & Lock in Spokane at the turn of the twentieth century. Teddy was holding his journal — a compilation of faded photographs and handwritten notes on the safes and vaults he worked on, most of them between the two world wars. Browsing that priceless piece of history was the one idle activity of mine that Mr. Corey would tolerate, even when the key machines needed cleaning and the floor needed sweeping.

"David," he whispered again, as Teddy handed me his father's journal. "We would like you to have this. There is no one left in the Corey family working in the industry, no one who would treasure this the way you will. Who knows, son, maybe you can even turn it into a book someday."

I was floored and grateful. And the following year, I did indeed turn it into a book. That journal became *Diary of a Safeman*, and its cover photo, showing the elder Corey and his son in their service vehicle — a Model T, no less — is one of my all-time favorites.

C. L. Corey and son, in their Model T service vehicle.
Source: Author photo.

Mr. Corey was getting tired. His eyelids were drooping, and his breathing was slowing. But I couldn't bring myself to leave; finality can be hard to face. I'd like to say we said our heartfelt goodbyes and shared one last memory, but we didn't. I simply stayed too long, and Mr. Corey fell asleep.

I was mortified, but Teddy assured me this was the preferred outcome. "Gene would not be able to say goodbye to you, David, without becoming very sad. In his condition, sadness is something he does not want to deal with. Truly, it's better this way."

Unable to speak, I rested my hand on his discolored, skeletal arm for a few seconds. Then I turned to hug Teddy and headed home in the darkness. Ten days later, she called to let me know that he had gone to sleep the night before and had not woken up. The book was still on his tray.

I think about Mr. Corey all the time. The image I carry is of him hunched over the workbench, peering through his Optivisor at marks on a key he is filing by hand. He was so good at it. I used to gently tease him about that dorky, glorified magnifying glass he wore on his head. Now, decades later, I have an Optivisor of my own, and it elicits an occasional giggle from my own kids.

I suppose it's true: the more things change, the more they stay the same.

THURSDAY

CHAPTER FIFTEEN

Morning Potpourri

Up several hours before dawn, I head straight to the computer with a large mug of dark roast, flavored with a dollop or ten of hazelnut creamer. Val sometimes asks if I'd like a little coffee with my creamer. I do love hazelnut. Without it, coffee tastes like swill, and, without coffee, I'd start each day even more sluggishly.

Yeah, I'm a slow starter, like a locomotive pulling out of the station. It usually takes a while before the words begin to flow from brain to fingertips. But today is a welcome exception. By the time the kids are ready for school, final versions of the articles for my safecracker magazine are finished, four days before deadline. This just might be a personal record.

The kids and I head to the van as the files are uploading to the Dropbox folder I share with the art department at National Publishing. We're circling the roundabout to enter the long driveway at the school when I notice the conversation behind me ratcheting up:

"Yes, it is!"

"No, it's not!"

I don't know what Aidan and Julian are disagreeing about, but a glance at my rearview mirror shows Claire grinning wide, eyes ping-ponging from brother to brother.

"It is!"

"It's *not!* A fifth-grader said it after lunch, and the recess teacher didn't even give him a timeout."

"So? She probably didn't hear him."

"Yes, she did! She gave him a timeout last week when he said the 'F' word!"

I'm starting to get the gist.

"Daaaaad, isn't *douchebag* a bad word? A cuss word?"

We coast into the parking lot and join the long line of cars waiting to drop children at the entrance. I tilt the rearview mirror down a little to make visual contact with all my passengers. Biting my lower lip to avoid smirking, I save the literal definition for later and address the more pressing question. We only have a minute, two tops.

"Well," I begin, Reagan-style, buying a couple of seconds to think. "It isn't as bad as the 'F' word or any of the other cuss words we've talked about before. But let's avoid it."

I steer the car a little closer to the curb, careful not to knock over any of the people crossing in front of us.

"Okay, Dad, but what does it mean?" asks somebody, followed immediately by another voice. "Yeah, what *is* it?"

I need more time. "Well, um, if somebody calls somebody else a douchebag, they are basically calling them an obnoxious jerk."

Three perplexed faces are evidence that my reply satisfied no one.

"Dad," says Julian in a pleading tone. "That's what you said the a-hole word means. What the heck? Do all cuss words mean the same thing?"

I tick through a few in my mind. They *do* seem to mean something sort of similar.

The van comes softly to a stop. I twist in my seat to face the congregation. "You guys, I haven't thought this through, but I think Jude is onto something. When someone gets mad and yells a bad word, they are expressing disapproval or frustration or anger. I don't think there is any precise meaning. It's sort of like shouting *grrrrrrrrrr*, only ruder. I guess many bad words do mean pretty much the same thing."

I hit the open button for the passenger side sliding door. "So, you're both sort of right," I continue. "Douchebag isn't quite a cuss word, but it's still bad. Definitely don't say it."

"Daddy, is it a compound word?" Claire asks as she steps out of the van.

Chuckling, I think for a second. Douchebag or douche bag? "Hmmm, I think it's two words, sweetie, but I'm not a hundred percent sure."

Aidan pulls the lever to shut the door. I prefer individual goodbyes, but my window of opportunity is closing quickly. I have just enough time to yell out, "Love you guys, see you this afternoon!" before the door clicks shut.

I'm still smiling as I head back to the house. I haven't thought deeply on this topic, but it does seem that swear words lack any sort of precise meaning. Weird.

I listen to a voicemail from Robert Coleman, a Diebold service technician based in Utah. Robert and I have worked together on several vault jobs in the Beehive State, but this is the most intriguing and challenging one yet.

The photo he just sent shows a full automatic bank vault. *A full automatic!* No dials, no handle, nothing. The vault door has a blank front except for the pressure system, which pushes the door in and out of its towering frame.

A full automatic bank vault is a monster to open; in fact, it's even a monster to close. At quitting time, bank personnel wind hours on each of the three tiny clocks in the timelock. They then cock the spring in the automatic boltwork actuator (the device that retracts all the bolts to unlock the door, and extends all the bolts to lock the door).

This may sound routine, but it's way out of the ordinary. The cocking tool itself is the size of a tire iron. It has to be strong enough to overcome the back pressure exerted by the gigantic flat spiral springs inside the automatic boltwork actuator that push and pull the boltwork.

Once the spring is cocked, bank employees push the multiton door into its frame and spin the captain's wheel on the pressure system to properly seat the door. When the door is all the way in, a small lever is depressed by contact with the frame. This triggers the boltwork actuator, which pushes the massive boltwork to the fully locked position.

On this particular evening, everything seemed to work like it normally does. But the following morning, when bank employees tried to draw the vault door out of its jamb, it wouldn't budge. Something had malfunctioned.

This is only the second full automatic bank vault I've been called to open. I feel a rush of adrenaline just thinking about it. Regular vault doors are fitted with dials and a handle; in the event of a lockout, these can be used to help diagnose the malfunction. Without them, all we can do is put an electronic listening device on the door. If the clocks are ticking, the lockout is probably due to an overwind. If that's the case, most banks will opt to let the time wind down rather than endure the expense and hassle of having the vault opened.

The bank has waited five days — the maximum on this particular timelock. Robert has used his listening device to confirm the absence of ticking. Troubleshooting is done, the bank is antsy. It's time to get serious.

I check outbound flights. I always take Delta from Portland to Salt Lake City (SLC), as they are the only carrier with nonstop service both ways. I book the ten-fifteen morning flight and the ten o'clock return that evening. That should be plenty of time.

I'm hoping to use miles to upgrade, but no go — first class is booked solid. This can happen on last-minute flights because the seats have already been opened up to Medallion members. Maybe I'll get lucky and move up the waitlist. After texting Robert my itinerary, I'm good to go for tomorrow.

I put my phone down, tilt back in my chair and close my eyes for a few seconds, and hear the double vibration of a text. It's from Allied Safe & Vault, where I worked for four years after saying goodbye to Mr. Corey and general locksmithing. The Portland Air National Guard is locked out of their armory vault door, and they need it opened immediately.

Truth be told, I'd rather stay home, work on my upcoming book on round door safes, and pick up the kids after school. But they need shoes, and we need food, so I text Allied that I'm en route.

I much prefer driving over flying to vault lockouts because I can take *all* of my tools rather than the skeleton set I fly with. My work van is always loaded for bear. That's often helpful, because tangling with these newer government vaults can be like rasslin' a grizzly.

CHAPTER SIXTEEN

The Air National Guard

AT THE AIR NATIONAL GUARD FACILITY, I'M QUICKLY ESCORTED DOWN a hallway and around a corner to the armory vault. I expect a Mosler, as they were the dominant provider of safes and vaults to the government for decades, before their sudden and shocking demise in 2001. Approaching the vault, though, I realize that it's not a Mosler. It's a Hamilton.

Hamilton is a major player in the banking world, and their bank vaults are everywhere. But their success has been more limited in the government market. I can count the number of Hamilton's government vaults that I've opened on one hand, without using pinky or thumb.

I'm giving the vault the once-over when my escort breaks the silence. "The guy on the phone said you're the best safecracker in the country."

I turn and smile. "That was very nice of him."

"Is it true?"

I shake my head. "No. There isn't a best safecracker any more than there's a best doctor or best lawyer. There are too many specialties and subspecialties, and no one is top dog at them all."

"What are your specialties?"

"I'm a general practitioner with a fondness for bank vaults and micro-drilling."

"What's micro-drilling?"

"Drilling tiny holes to open safes and vaults."

"How tiny?"

"The diameter of an uncooked spaghetti noodle."

"Wow. Are you going to micro-drill this one? Is this a bank vault?"

"No, this is a government vault, and a very good one, built to spec with excellent hardplate. Micro-drilling isn't an option."

He nods. Satisfied, he backs away to let me work.

A simple rule that guides professionals: open the lock, and the safe (or vault) will follow. This strategy works when the problem is a lost combo

The Hamilton armory vault.
Source: Author photo.

or a lock malfunction. It fails when the problem is in the boltwork rather than the lock.

Today's problem is a little tricky to diagnose. The Air Guard has the code to the X-09 electronic safe lock, which was also custom-made for the government. Once the code is entered, the dial comes to a stop, like it should, but the vault's handle won't turn. I enter the code again, holding light pressure on the handle as I turn the dial to the stop position. If the lock opens, the handle will move, at least a little. But it doesn't move, not even when the dial comes to a complete stop. This strongly suggests that the lock itself is not fully unlocking.

There are three common problems consistent with this behavior, but I run a diagnostic test and narrow it down to two. Fortunately, both possibilities have the same drill point.

My escort walks over as I'm removing the hybrid keypad/dial assembly and politely asks my opinion of these newfangled electronic locks. He points out that other armory doors use "regular" locks and asks if the electronic locks "crap out" more frequently than their mechanical counterparts.

"Yes, they do," I reply. "And just so you know, an electronic lock is not required on an armory vault. Would you like me to put a regular old spindial, mechanical lock on once we get the vault open?"

He nods. "Good idea. But let me make a call and confirm before we commit to that."

I return the nod and turn back to the vault door, but his curiosity isn't satisfied. "Well, when *is* an electronic lock required?"

I can't help but smile. "In its infinite wisdom, our government decreed that expensive electronic locks need only be retrofitted on containers housing confidential information, such as file cabinets, map and plan safes, and the like. Weapons safes and armory vaults are exempt."

He tilts his head. "Do I detect a little sarcasm? I gather that you think electronic locks should be used on *all* containers?"

I answer truthfully. "Well, yes and no."

"What do you mean?"

I turn to obliquely face my escort. "Yes to the sarcasm. But no, I don't think expensive electronic locks should be used on all or even most containers. In my view, the retrofit program is largely a waste of taxpayer dollars."

He is a little taken aback and reasonably asks, "Why?"

I rattle off four long-memorized reasons. "First, the cost. These special electronic locks are more than ten times as expensive as the original mechanical locks, which are already paid for. Second, electronic locks fail at a much higher rate and don't last nearly as long — we are talking years versus decades. Third, the security concerns used to justify the whole enterprise are dubious at best. Fourth, and finally, low-tech vulnerabilities were brought to the government's attention more than twenty years ago, and yet they never unapproved the locks."

"Seriously?" he replies and sits down. "Why not?"

I shrug. "I truly don't know."

"Well," he wonders, "just how expensive are they?"

I point to the keypad. "My cost on an X-09 through a distributor is just under a grand. But you guys can buy them off the Federal Supply Schedule for a little over seven hundred bucks apiece."

His eyes are pie plates. "For real?"

"I know it sounds loony, but yes. This is a lock with a price tag only Uncle Sam could love."

I begin measuring out the drill point, drawing perpendicular lines on the blue tape and marking their intersection with a Sharpie dot. A slow burn begins to form as I think back to how this sad state of affairs came to be.

There was a nasty episode involving me, the lock manufacturer, and one of my closest friends in the industry. For many years after, I was legally gagged and couldn't talk about it. But there was an expiration date. It has passed.

CHAPTER SEVENTEEN

The Great Lock Debacle

THE U.S. GOVERNMENT HAS MANY SECRETS. THE MOST SENSITIVE OF these are kept in special safes made for the government, to specifications set by the government. The vulnerable point of every safe is its lock, whether mechanical with tumblers and a dial, or electronic with circuits and a keypad. Open the lock and you open the safe. Hence the government's long-standing desire for combination locks that are virtually manipulation-proof.

In the late 1980s, a Bigwig from deep inside the safe and lock industry scared the bejesus out of some key representatives from government agencies concerned with physical security. He showed them just how vulnerable their safes were to high-tech methods that leave little or no evidence of a compromise. He introduced them to sophisticated autodialers, computer-enhanced neutralization techniques, advanced imaging technology, and more.

The agency reps were understandably alarmed. The image of a foreign spy sneaking into a secure facility and surreptitiously opening a government safe to steal our national secrets was unsettling. The reps soon presented the General Services Administration (GSA) with a list of performance features their fantasy lock would possess. Features, by the way, that were suggested by Bigwig.

To the amazement of nearly everyone involved, there was no negotiation and virtually no review and comment. The wishlist was drafted into Federal Specification FF-L-2740. And what a stunning coincidence: the only lock with any chance of meeting the requirements of FF-L-2740 was Bigwig's breadboard prototype. It was sold to a wealthy businessman who assembled a team to refine the design for mass production. Bigwig would be paid on a royalty schedule, based on the number of locks sold.

Bigwig's concept lock was soon actualized, and the most technologically advanced combination lock ever made was given a model name straight out of James Bond. And when the X-07 passed all the tests at the government's lab, it became the *only* approved lock for use in GSA containers — excepting armory vaults and weapons containers, which can be fitted with mechanical combination locks.

The X-07 was introduced with much hype and hoopla. The manufacturer appeared at the Associated Locksmiths of America convention to hand out X-07 shirts with bold-face text across the chest: "THE ONLY WAY TO PICK THEM IS TO CHOOSE THEM." The accompanying brochure bragged that it was "impervious to external attack."

The passage of FF-L-2740 and subsequent approval of the X-07 shook up the industry. Most shaken was Sargent and Greenleaf (S&G), whose combination locks had been the gold standard for safes and vaults going back more than a century. Without an approved lock of their own, S&G's sales to the government fell to nearly nothing. Frustrated and feeling a bit blindsided, they sicced their own testing engineer on the X-07 with instructions to develop a defeat method. He quickly succeeded, popping the lock open in twelve minutes using nothing but a bent wire.

This was demonstrated live to a congressional delegation, which greatly embarrassed the technicians from the government lab. They had certified the lock as being resistant to covert entry attacks for a minimum of thirty man-minutes (as required by the specification). The manufacturer was forced to suspend production. But instead of rescinding FF-L-2740, which would have put S&G back in the game, the government helped the manufacturer apply a Rube Goldberg fix. The bandaged X-07 was back in production three short months later, still the only approved electronic lock for use in GSA containers.

S&G was unaccustomed to playing second fiddle. Frustrated by this turn of events, they set out to find a lock-defeat artist who could absolutely devastate the X-07. They assumed that if no fix were possible, it would *have* to be unapproved.

S&G had the good sense to approach my old friend, Mike Madden. He ran the locksmith shop at Lawrence Livermore National Laboratory during the day and freelanced after-hours. Although Mike wasn't particularly well known among the rank and file, he had a stellar reputation among manufacturers. Most of them had employed his services at one time or another, either to test their own locks or to devise defeat methods for competitors' products.

Mike signed on, motivated by both the nice contract and the unique challenge. Nothing spurs innovation among elite keylock pickers and combination lock manipulators like a manufacturer's claim of invulnerability. It's blood in the water. S&G also persuaded him to sign a two-year nondisclosure agreement during which they were forging ahead to finish their own lock.

S&G's plan was tactically sound. First, expose a fatal vulnerability in the X-07 to get it unapproved and therefore removed from the Qualified Products List (QPL). Then they could unveil their own lock, get it approved and on the QPL, and thereby replace the X-07 manufacturer as the only provider of electronic safe locks to the government.

Mike went to work. He didn't waste any time on the high-tech stuff. He surmised correctly that the lab techs had thoroughly tested the lock for resistance to surreptitious attacks from X-ray, backscattering, autodialing, resonant vibration, high voltage, shock, an electromagnetic pulse from a simulated high-altitude nuclear blast, and more.

Mike knew his best chance was to go totally *low*-tech. He focused all his efforts on the lock's Achilles heel: the lazy cam on the back side of the stepper motor. Move this tiny little cam, and the lock opens. He spent several days guiding compressed air and tiny tools into the lock. He was able to move the lazy cam, but it was unpredictable and intermittent. He wanted something more reliable.

The following morning, while squirting toothpaste onto his toothbrush, Mike had the inspiration to try liquids. The idea was to inject a substance down the spindle hole into the lock. He needed a liquid that was thin enough to squeeze through small cracks, but viscous enough to snag and rotate the lazy cam.

Over the next week, Mike experimented with a wide variety of substances, including corn syrup, molasses, shaving cream, toothpaste, denture cream, hair gel, insulation foam, grease, Fix-A-Flat, and a bunch of spray adhesives. Almost all of them were eventually able to move the lazy cam, but one particular product was the clear champion.

Marketed by BF Goodrich as Plastilock S900, this water-thin spray adhesive flooded through the tiny gaps into the lock, where it soon got tacky. Mike began popping open X-07s in five minutes or less, and it worked every time. Less than two weeks into the project, satisfied with his results, Mike ended testing and submitted his findings to S&G.

Surprised to hear from him so quickly, and a little skeptical that a can of glue could actually be the nemesis of their nemesis, S&G flew their engineer out to meet with Mike and document the procedure. In the middle of his garage, Mike squirted a little Plastilock S900 into ten brand-new X-07s. Every single lock was opened in an average time just under five minutes.

The race was on. S&G now had a fast and reliable defeat for which there did not appear to be an easy fix. They just needed to finish their own lock, get it approved, and take this one-two punch to Uncle Sam.

S&G named their lock the 9000. Perhaps it was a coincidence that the number they chose simply added a zero to the glue that defeated the X-07. Or maybe it was intended to tweak their competitor. If so, the tweak failed, because the 9000 never made it through testing.

The government lab failed the lock for the most infuriating of all reasons: *just because*. Just because of this or that theoretical possibility. Never mind that no one at the lab could think of any way to test their theories in the real world. Five times in succession, the bruised egos at the same lab that passed the X-07 used hypothetical what-ifs to fail the 9000.

It became clear to the brass at S&G that their lock was no-way, no-how going to be approved. So, with two years and nearly two million dollars in research and development circling the drain, S&G reluctantly pulled the plug on the project. They were forced to give up on government sales and shift their focus to the commercial side of the safe lock business. Without a lock of their own in the game, S&G was no longer interested in the X-07 defeat. They paid Mike and moved on.

Other than Mike, no one outside S&G yet knew about the glue trick. That was about to change. The day the nondisclosure agreement expired,

Mike called me and spilled the beans. He had dropped little hints along the way, so I suspected he had been working on something exciting or important. Still, I was surprised to learn that he had taken down the X-07, and astonished by S&G's decision to walk away. Most industry insiders had assumed the 9000 would eventually be approved.

As a professional safecracker, I was fascinated by this story, and, as a writer, I quickly realized that my old friend had just handed me the biggest scoop of my career. But I needed tangible proof. So Mike drove up to my home in Washington State, where we spent all day gluing locks open and video-documenting the process.

I even made a minor contribution to the method. The original defeat required a two-minute wait after the watery glue was injected into the lock, for it to thicken and get tacky. I suggested we skip the wait, and rotate the lock's spindle in a cordless drill to see if the heat generated inside the lock would speed up evaporation and result in timelier tackiness. It did. By day's end, we were popping locks open in under four minutes, in a freezing garage, in the middle of winter.

As surely as night follows day, claims of invulnerability by lock makers are eventually proven false. After so many humiliations over the past two centuries, you would think that the manufacturers would notice the pattern and button the braggadocio.

I couldn't wait to break the news to the industry via my safecracker magazine. I shared Mike's glue trick with a few of my more discreet friends, among them Don Spenard, who insisted that we hold the presses, as he had a "killer idea" for a photograph to accompany the article.

Don had snagged one of the X-07 promotional shirts at the recent convention. He wouldn't tell me exactly what his idea was, but only that it involved the shirt, which was already on its way to Mike. A week later Mike called, cackling. He, too, refused to tell me anything, other than to tease that the U.S. Postal Service was in possession of a glossy print that was going to make my day.

The reward was worth the wait. I was proofing the article when the envelope was delivered. I sat down and took a large gulp from my favorite mug before pulling out the photo, and darn near sprayed Starbucks blend all over the top of my desk.

The photo highlighted Don's comedic capacity for audacity. In it, Mike was wearing the X-07 shirt, but a single word of the original text had been altered. Don had crossed out "choose" and Sharpie-marked "glue" in its place. The caption now read: THE ONLY WAY TO PICK THEM IS TO **GLUE** THEM.

Priceless.

At the time, it didn't occur to me that the manufacturer of the X-07 might not find the photo as funny as I did. It probably should've but didn't. In any event, the photo would feature prominently in the article, which I sent off to my publisher the following afternoon.

Mike Madden, sporting the infamous "GLUE" shirt.
Credit: Mike Madden.

The first half was an editorial in which I shared my not-so-humble opinion about the retrofit program. My question was simple. Does it make sense to remove hundreds of thousands of long-lasting mechanical locks that we have already paid for and replace them with new locks of questionable reliability that will cost exponentially more?

I chastised: the GSA for turning Bigwig's wish list into a FedSpec that favored his concept lock with so little review and comment; the government lab for not utilizing first-tier technicians from outside the lab when testing for covert vulnerabilities; and the X-07's manufacturer for not seriously addressing the lazy cam vulnerability.

I also speculated on the relationship between certain Kentucky politicians and the manufacturer. I suggested that it might be a crass example of crony capitalism, perhaps bordering on corruption. There's a reason that S&G demonstrated the wire defeat to a congressional delegation from Indiana, rather than one from their home state of Kentucky.

The second half of the long article was a photo essay divulging precise details of the defeat method. In the center of the middle page was the photo of Mike Madden, proudly wearing the Sharpie-marked shirt and a devilish grin.

My regular routine is to look over proof pages as soon as they arrive, make my edits, and hand them right back to the production team for final preparations before going to press. This one was different. It was the longest, hardest-hitting piece I'd ever written. Given the nature of the claims made in the article, it seemed unwise to turn everything around so quickly. I wanted to make sure my facts were right and allow everyone involved the opportunity to chime in so that no one would feel ambushed.

So, I faxed drafts of the article to Mike, to the manufacturer of the X-07, and to Bigwig, explicitly inviting comment and criticism. Yes, I did.

The next morning the proverbial poop hit the fan and sprayed in every direction. I awoke to messages from Mike, my publisher, Bigwig, and the manufacturer of the X-07. The manufacturer's message was crystal clear: the article was defamatory and libelous. Moreover, the t-shirt photo wasn't merely unfunny, it was a blasphemous defacement of their intellectual property. They threatened to sue if the article were distributed.

I didn't doubt it — they already had a reputation for having a hair-trigger on the litigation bazooka.

Mike Madden immediately made an appointment with his attorney.

My publisher spoke to their First Amendment specialist and was advised to avoid provoking a fight. I agreed to the removal of the publisher's name from the article and its mailing envelope; we arranged for it to go out as a special edition from me directly to the membership.

Bigwig was an industry titan from a family of titans going back several generations. He stood to make millions of dollars in royalties from the manufacturer who had turned his concept lock into the X-07. If the article was distributed and the lock became unapproved, it would cost him dearly. Yet his primary concern when we spoke was the suggestion of possible impropriety between Kentucky politicians and the manufacturer.

He was right that I had no substantial evidence, just a connect-the-dots argument. I liked the argument, which was nested in an editorial rather than a news story, but I didn't want a peripheral issue to take center stage. So I agreed to excise the offending comments. The real punch in the article came not from my puffed-up pontifications but from Mike's ingenious glue trick. The deletion was painless.

Mike called right before supper, dejected and apologetic. The meeting with his attorney hadn't gone well. They decided the best option was to minimize his exposure and avoid a potentially expensive court battle. "I won't tell you not to publish the article, Dave, but I'd prefer to avoid a lawsuit."

I understood completely. I was bummed but did my best not to let on, while I racked my brain for a solution that would satisfy everyone. There just didn't seem to be one.

I awoke the following morning to a message from the president of the lock manufacturer, asking me to call him back ASAP. He called again during breakfast. Chewing a mouthful of Grape-Nuts bathed in extra-pulpy orange juice, I listened as he began to leave a message.

I wasn't prepared to have this conversation. I wasn't ready to throw in the towel but didn't know how to avoid it. I didn't know what to say because I didn't yet know what I thought. I don't often act impulsively, but I reached over and picked up the phone.

The conversation started pleasantly enough. He was calm and cordial for the duration of the obligatory small talk. Then he filled up his lungs and launched into a tirade. He slammed me for "yellow journalism" and "that crony capitalism crap," before accusing Mike and me of being insensitive to national security and calling the glue-defeat method "complete bullshit." By the time he took a breath, he was highly agitated but still somewhat rational.

I disagreed with him strongly about national security. The real threat, I maintained, has never been from the outsider — the foreign spy with an invisibility cloak who can pick any lock and crack any safe. The *insider* is who we should worry about — the person who already has access and is compromised by ideology, greed, or sex. Think Aldrich Ames, Robert Hanssen, John Walker, and, more recently, Chelsea Manning and Edward Snowden. *A better lock won't stop an insider who already has the combination.*

His reply was a classic bluff. He insisted that he was privy to important information about which I was ignorant. I needed to accept that the outsider threat was absolutely real. In other words, if I knew what he knew, I would stand foursquare behind FF-L-2740 and the X-07.

I paused and struggled with how to respond to such tall gall.

The empty air irritated him. "Hello? Dave? You still there?"

"Yeah, I'm still here," I deadpanned. My decision was made. Rather than challenging his claim, I accepted it as provisionally true and pointed out the obvious. The X-07 didn't even come close to meeting the requirements of FF-L-2740. The spec required a lock to resist covert attack for a minimum of thirty man-minutes, yet Mike's glue trick was popping them open in less than four. That left a huge twenty-six-minute gap — more than enough to get the X-07 unapproved and booted off the QPL.

But no one inside the government knew anything about this. Yet. And my incensed interlocutor hoped to keep it that way, at least until the next zillion-dollar contract was signed, to eventually put an X-07 on every personnel file and shoebox owned by the government.

He immediately went on the offensive and again attacked the glue-defeat method as a joke. He threatened legal action for the second or third time. I was starting to feel irked, so I asked if he were a closet masochist who wanted to be embarrassed in court by a jester spraying "joke" glue into his lock and popping it open in mere minutes. Sensing an imminent and unnecessary escalation, I tried to soothe things by injecting a little levity. I asked if he had chuckled the first time he laid eyes on the photograph of Mike wearing the altered X-07 shirt.

Up to this point, the exchange had been a contentious but semicivil session of verbal thrust and parry, punctuated with a few sharp but relatively restrained zingers. It now went nuclear as he morphed into a mushroom cloud over mention of Mike's smirking snapshot. I had to hold the phone at arm's length to escape the vulgar, high-decibel freak-out. It was a timely reminder that one man's humor is another man's insult, and that *touché* is not in everyone's jousting vocabulary.

I resisted the urge to hang up and waited instead for the shouting and swearing to subside. "Are you done?" I asked, modulating my voice and pretending to be more amused than annoyed. He responded by slamming the phone down and then calling back a minute later. This time I felt no impulse to answer.

I sat there, a little shell-shocked. The more I processed what had happened, the more determined I became to publish the article, mail it to my merry little band of safecrackers, and find a way to protect Mike. And then, out of the blue, a solution occurred to me. We would blur out Mike's face in the photograph and call him "Mr. X."

Ta-da!

Thrilled, I called Mike, got his go-ahead, and instructed National Publishing to make the changes and distribute the article post haste. Batten the hatches and man the torpedoes. Mushroom-cloud man had made it personal.

Mike called a few days later from his lawyer's office. He was panicked. Obscuring his name and face might have protected him from being named in a lawsuit, but only if the relevant parties didn't know who he was. Since a draft of the article had already circulated, his identity was known. He could and would be named.

Duh. I should have known that, or at least run the idea by my lawyer wife.

Mike was told that he would almost certainly win, that such a lawsuit might even be deemed frivolous or malicious. But there was no guarantee that the presiding judge would award attorney fees. He asked me to withdraw the article.

I immediately called Marc Goldberg, my boss at the publishing house. Marc had seen the pallet of magazines sitting on the dock awaiting pick up earlier that morning before he drove across town to a meeting. He would find out if the pallet was still there and call me right back.

Those two minutes were among the longest of my life. I picked up on the first ring. Marc had good news. The magazines were still there, and yes, they could be intercepted. Was I sure that was what I wanted? Without a doubt.

Yeah, I wanted to hit back at a company I viewed as a schoolyard bully, but it wasn't worth the potential cost to my friend. And so, two thousand copies of the longest and best industry article I had ever written were shredded and promptly dropped into the dumpster.

Mike breathed a sigh of relief. I, too, was relieved, but also peeved. It seemed wrong that a company could suppress accurate information about their product by threatening to sue anybody who dared whisper the truth. But there was nothing I could do. After weeks of excitement and intrigue, it was suddenly over.

Or was it?

Later that afternoon, mushroom-cloud man left another message, asking me in an eerily calm voice to please call him back.

I was unable to think of a single reason to return his call. I had nothing to say, and I didn't want to give him the satisfaction of knowing he had won. I pondered calling Mike and my publisher and asking them not to answer or return his calls. He would be out of the information loop, at least for a while. That prospect, admittedly petty, made me smile.

Then a much better idea percolated to the fore. True, the article was dead. It was never going to be in the hands of my readers. *But mushroom-cloud man didn't know it yet.* Hmmm, I wonder if he would be interested in buying my little masterpiece?

I don't have the poker face to do something like that in person. I turn beet red. But I can and did pull it off over the phone. I quickly sold mushroom-cloud man the copyright to an article that was never going to be distributed, that in fact had just been destroyed. It was easy, which probably means I left a pile of money on the table. But it felt good to get something for my dumpstered efforts.

As for my dear friend Mike Madden, the story did not end at the dumpster. When the DoD got wind of Mike's glue trick, they borrowed him from the Department of Energy (DoE) to test locks at the government lab. The manufacturer had ceased production of the X-07 and introduced the X-08, which Mike defeated almost as quickly, using a modified version of the glue trick.

Mike expected both the X-07 and the X-08 to be unapproved and removed from the QPL. It should have been an easy call by the powers that be, since the locks were providing a mere three or four minutes of resistance to covert attack, rather than the thirty minutes required by FF-L-2740. Inexplicably, neither lock was removed.

Mike eventually went to his congressman, Gary Condit, and divulged the details of this debacle. Condit was flabbergasted and sent a stern letter to the GSA, demanding answers to a host of unaddressed questions. Unfortunately, Condit became embroiled in the murder investigation of Chandra Levy soon after and was voted out of office.

Meanwhile, the disaster known as the X-08 — far and away the most failure-prone safe lock of all time — was redesigned and released as the X-09. Similar looking but better built, it incorporated the changes Mike had suggested way back when the X-07 was initially defeated. The DoD again borrowed Mike from the DoE, and he went right to work on the X-09.

He spent the first morning examining the lock. The lazy cam he had targeted before was gone, but he had a few promising ideas for other attacks. Well, as (bad) luck would have it, word of Mike going to Congress had filtered down from the GSA to the DoD and from the DoD to the head honcho at the government's testing lab.

When he returned from lunch, Mike was accosted by Honcho, who barked at him about the importance of chain of command and ordered him never to do anything like that again. Mike smiled and pointed out several things. First, the DoD's beloved chain of command didn't apply, because he was a DoE employee on courtesy loan. Second, they *had* tried chain of command over and over, and it hadn't worked — both locks were *still* approved. Lastly, Mike confessed that he couldn't tell whether the root problem was stupidity, incompetence, corruption, or some combination of them all. Whatever the case, it made his work futile, an utter waste of time and taxpayer dollars. No matter how fast he defeated the locks, they were never unapproved.

"What's the point?" Mike asked. "Why are we doing all this work, if the locks we defeat remain approved?" Honcho shrugged his shoulders and insisted that Mike agree to respect chain of command before going back to work.

At that moment, my old friend took a deep breath and decided to stop pissing into the wind. He walked back to his workstation and gathered his tools. After saying goodbye to his friends and colleagues there, he got into his car and drove three hundred miles back to Livermore, where he tried to finish out his career at the DoE quietly, under the radar.

Alas, a peaceful path to retirement wasn't in the cards. Someone with some serious clout decided to make him the target of an investigation. The claim being pursued was that Mike hadn't figured out the glue method himself, but had stolen the idea from someone else or some other project at Lawrence Livermore National Laboratory, and was therefore guilty of misappropriating government resources.

When the first investigation found nothing and exonerated Mike, it became clear that he was in the crosshairs of someone very high up, because a second investigation commenced immediately. This time, three DoE investigators and two lawyers from the DoE's Office of General Counsel were dispatched to Livermore to bend Mike over for yet another series of probes.

The second investigation dragged on for nearly a year before yielding the same result as the first. There was no evidence that the glue trick had originated anywhere other than inside the mischievously magical mind of Mike Madden.

By this time, Mike was pretty ticked off, to put it mildly. He wanted to know who had put him through two investigations and why, so he hired an attorney to chase the paper trail. First stop was a brick wall. The glue-defeat method, the article about it, the investigations themselves, and all relevant documents had conveniently been classified, leaving no accessible trail to chase.

This forced Mike to sue his own employer to declassify the glue-defeat method, the article about it, and related documents. It was an easy win. Since there existed no grounds for suspicion, much less probable cause for an investigation, and since the government had classified things far outside their proper purview, Mike prevailed — nearly everything was declassified.

Although Mike's lawyer was ultimately unable to unmask the identity of the prime mover, this much was clear: either someone high up at the DoE went after Mike for their own petty reasons or they went after Mike on behalf of someone else's petty reasons. Either way, it reeked.

To demonstrate how disingenuous all the hysteria over "national security" was and still is, consider this: the X-07 and X-08 are *still* approved! An unknown number of GSA containers still have locks on them that can be opened with Mike's glue trick.[1] The others might be equally vulnerable, but we don't know, since Mike didn't finish his work on the X-09, and never got started on the newer X-10. After all, *what's the point?*

All of the relevant parties were informed, from the government testing lab, to the GSA, to Congress. And no one but Gary Condit ever did a thing. The grave concerns that gave rise to Federal Specification FF-L-2740 and the eventual purchase of more than one *million* locks by our government seem to have evaporated like the money that was so casually spent on them.

Chapter Eighteen

Opening the Armory Vault

As I prepare to drill the X-09 on this armory vault for the Air Guard, something doesn't look right. This is a tough door. Did I triple-check the drill point? I did not, and the hash marks I've drawn on blue painter's tape are obviously not in the right place. I slow down and re-mark the drill point in a spot that will permit the insertion of a stiff wire to probe the lock-bolt.

I take out my phone and open WWF. It's still Val's turn. *Slowpoke!*

Back to the vault. It's been here for all of three years, which tells you exactly how long this thousand-dollar lock has lasted. I triple-check the location of the Sharpie dot, and it's perfect.

I fire up the Fein, drill through the dot, then the door's outer skin, and stop at the hardplate. This is a "red-label" vault door, which means the hardplate is hard as hell. You see, back about the same time Bigwig was selling his concept lock to the startup manufacturer, he also convinced the government that their safes and vaults were too easy to drill, that the container specifications needed to be upgraded to require better-quality hardplate. You can guess who had just developed some excellent (and expensive) hardplate, ready for sale as soon as the next omnibus bill with a juicy rider was passed by Congress.

Bigwig has done quite well. I don't begrudge him his gazillions. He is a smart and savvy businessman, persuading the government of a need, and then filling that need with heaps of product. I just wish common sense were a little less uncommon, especially when it comes to Uncle Sam's handling of our money.

It's time to inspect the hardplate, so I open the pistol case and grab my favorite scope. As expected, I see the dreaded texture on the hardplate. The texture is from a welding process patented by Bigwig. He starts with an abrasive-resistant plate, and then runs beads back and forth from hard-facing rod filled with carbide pellets that flow onto the plate as particles of varying sizes. It's some of the most drill-resistant hardplate ever made.

I lean into my lever rig with a 1/4″ bit and am pleasantly surprised by the rate of progress. Today I'm either on my game or really lucky. It usually takes me an hour or two to drill through this nasty stuff, but this time I sail through it in just under thirty minutes. Fantastic.

With the hardplate penetrated, I switch to a 1/4″ hole saw to finish the hole through the lock case. Damage to a lock's internals is always a concern because many of the targets we aim for are small, fragile, and easy to ding with a hot drill bit.

I'm not worried, though. My bulls-eye is on the lock-bolt, the beefiest component in any safe lock. It's a little larger than that big slab of butter you slice off the cube to put on your pancakes. For a safecracker, a target that size is like center mass to a sharpshooter. I won't miss.

Once the hole saw breaks through into the lock, I take my finger off the trigger and lay the drill motor down. My arthroscope slides through the drilled hole and into the X-09 for a look inside. My sigh of relief is almost audible. Nothing but chubby, glorious lock-bolt!

I chuck up a 3/32″ drill bit and dimple the lock-bolt at the extreme left side of the hole. Into the dimple goes my stiff wire, and I probe from left to right, until the lock-bolt succumbs to the tug of wire and slides to full retraction.

Rotating the vault's handle, I hear noise at my six and glance back at a gaggle of onlookers who have been waiting to gain access to their munitions. I pull the armory door open, letting it sweep my Beemer case and tools along the wall so no one will trip.

Cheers of "Yes!" and "All right!" fill the air as airmen in camouflage battledress file into the large room to access the weapons lockers lining the walls. They are late for drills and I for supper.

CHAPTER NINETEEN

Dad and Morty

HEADING NORTH TOWARD THE BRIDGE ACROSS THE MIGHTY COLUMBIA, I'm listening to a podcast when my phone rings. Glancing at the screen, I see that it's my father. Whoa. He doesn't call very often, which heightens my concern, so I answer immediately.

"Hi Pops. Is everything okay?"

Dad would rather walk on hot coals than engage in lengthy I'm okay/you're okay white noise posing as meaningful social intercourse. He comes right to the point. "Of course everything's okay, except that I haven't been able to find Morty's second autobiography. Did you borrow it by chance?"

"Um, no." And then I remember. "But I might know where it is. I saw it on the end table next to your recliner when I dropped off mom's Mason jars over the weekend. I have always liked that book, Dad, and I browsed it for a few minutes while you were lying down. I may have absent-mindedly walked into your office and put it back in your library. I might also have misfiled it. Do you want to check?"

"I already did, but I'll look again. Hang on."

Morty is Mortimer Adler, the American philosopher who wrote his first autobiography, *Philosopher at Large*, at age seventy-five. He lived long enough to write *A Second Look in the Rearview Mirror* at ninety.[1]

My first Adler book was a Christmas gift from my father, a truck driver with an insatiable thirst for knowledge. He plucked *Aristotle for Everybody* out of Waldenbooks' bargain bin in the late 1970s and couldn't put it down. It started us on a learning journey from which there is no graduation, no finish line, and no escape. Philosophy's allure is as exhilarating as it is exhausting.

Dad comes back on the line, clearly in a better mood. "Found it. Thanks, son."

"That's great. Was it with the 'A' books?"

"No, it was with the 'M' books."

"Sorry, Pops."

"No big deal, David. Just glad it's here. Well, I'm going to read now. Bye."

Mortimer Adler was a frequent guest on *Firing Line*, and all of his appearances were gems, partly due to the conversational contrast. Buckley, the host, was a master at language adornments and elevated discourse, while Adler eschewed big words, preferring effective communication over sesquipedalian elegance. Their exchanges on a variety of topics are extemporaneous masterpieces, and I feel fortunate to have witnessed (and recorded) most of them.[2]

One of my favorite *Firing Line* episodes was devoted to Adler's book, *How to Think about God*. His explanations were clear and concise, filled with ordinary words and common sense. I immediately went out and bought it in hardcover. Dad and I read and discussed it at length over the next month.

I did some (pre-Internet) sleuthing, tracked down an address for Adler and fired off a lengthy letter containing several dozen questions. His brief reply was as surprising as it was scary. He invited me to spend an afternoon at his Institute for Philosophical Research in downtown Chicago for an in-depth, face-to-face discussion.

My publisher's office was also in Chicago, so, on the next trip east, I came in a day early for an intimidating encounter with a truly prodigious mind.

Adler was surprisingly friendly and engaging. But when the conversation turned to his book and my questions, his demeanor changed palpably. His head tilted back, and his dark, serious eyes narrowed as they tractor-beam locked on mine across the desk. I had Mortimer Adler's undivided attention, and he had mine.

An attentive listener who practiced what he preached — Adler's *How To Speak, How To Listen* should be mandatory reading at every high school — he leaned forward to ensure that nary a syllable was missed. I would state each objection, then he would rephrase it into a far more precise and powerful collection of words and confirm that that was what I actually meant. I would nod. Then, with merciless, surgical precision, he proceeded to slice each and every one of them to pieces. Ditto for all of my follow-ups.

I'm a debater at heart, but it felt like I had come to a gunfight armed with Nerf bullets.

Early that evening, I left Adler's office not only with my list of questions answered and obliterated, but also with an unexpected invitation to attend his week-long spring and summer seminars at The Aspen Institute as his comped guest. Although he never offered an explanation for this kind offer, I suspect he liked (what I hope he perceived as) my polite tenacity. And he got a kick out of my profession. "Dave is a professional *safe*cracker!" he would announce with a chuckle, as I shook hands with his friends and acquaintances.

I was a frequent participant in Aspen for most of the next decade. We covered a range of Adler's favorite topics: the existence of God, free will, the mind/body problem, foundational issues in moral theory, and more. It was an intellectual smorgasbord attended by some very smart people, and I felt privileged to be in their presence.

Mortimer Adler did two important things for me, and I imagine his seminar attendees and the readers of his many books would say the same. First and foremost, he made complex issues accessible to anyone who can read. The importance of that will be evident to anyone who has picked up a primary source in just about any academic discipline outside their area of expertise. We've all slogged through a wall of technical jargon trying to understand what in the hell an author is trying to say. There are no such walls in any of Adler's post-1976 books. He jettisoned the jargon and composed short paragraphs with short, clear sentences.

The second thing flowed directly from the first. Since his words, both spoken and written, are models of exposition and can be understood by ordinary people, they tend to deepen rather than diminish interest in the weightiest issues human beings can confront. My father and I were hooked, as were many others who might have shied away, due to the ponderous, nearly impenetrable prose that fills so many works of philosophy.

The one and only time I visited Mortimer's vacation home in Aspen was the evening after his retirement seminar. It was a packed house. While

I was fortunate enough to be seated next to Adler at the Happiness Banquet, c. 1993.
Source: Author photo.

waiting to pay my respects, I spied memorabilia on a long table across the wall from his elevator. (Yes, an elevator!)

I sauntered over for a closer look. Among the many items on the table was a large photo in a wooden frame that captured my gaze and raised goosebumps of nostalgia all down my neck. Other than my parents and Mr. Corey, the two people who had the most influence on my life were philosopher Mortimer Adler and fictional safecracker Alexander Mundy, played by Robert Wagner. I would never have wandered into Mr. Corey's lock shop without Wagner, and I would never have become interested in philosophy without Adler. And there they were, sitting side by side, smiling for the camera.

I glanced at that photo a dozen times or more during that memorable evening and have thought about it many times since. I wish I had had the nerve to ask for a copy to hang on the wall in my office.

The juxtaposition is jarring. And yet fitting. Safecracking and philosophy — different paths to unlocking life's secrets and exposing her treasures. One became my vocation, the other my avocation.

A beautiful bounty indeed.

FRIDAY

CHAPTER TWENTY

Pinky Swear

AIDAN AND JULIAN HOP OUT AT THE CURB, HOLLER PRETTY MUCH IN unison, "Bye Dad!" and speed-walk toward the entrance to the school. Claire leans over the console, puts her hand on my shoulder, and reminds me of the big upcoming event — her dance recital tomorrow morning. "You promise to be back for my performance, right Daddy?"

"Sweetie, I promise to do my *best* to make it home in time. You know I wouldn't want to miss your recital!"

I look over, expecting acquiescence, but Claire isn't happy. "That's not a very good promise, Daddy. Can't you just promise that you'll *be* there?"

I should have anticipated this objection and formulated a counter beforehand. But I didn't, and so I say the first thing that comes to mind, in the third person to give it an air of authority. "Well, Daddy can't control if a flight gets delayed or canceled. Remember what happened a few months ago?" (I spent the night in O'Hare's Concourse C during a snowstorm.)

"Okay Dad," she says, gears already turning a little faster than mine. "So you promise to be home if your plane ride isn't late?"

I'm cornered and know it. I feel like exhaling a Junie B. Jones huffy breath, but the sheer ridiculousness of a second-grader checkmating a reasonably intelligent adult in two swift moves forces a chuckle instead. "Yes, sweetie, I promise."

Her little hand shoots out in front of my face. It's balled into a loose fist with her little finger curled into a "J."

"Pinky swear?" To Claire, this is the highest, most solemn form of a promise, which means it is to me, as well.

There's no going back now. Retreat not an option, I curl my little finger around hers to seal the deal. "Pinky swear."

Claire beams. Not a victory smile; a delighted, *happy* smile. She gives me a super-squeeze hug and hits the lever to open the side door. As it's opening, she turns and says, "Daddy, do you know why Waldo wears stripes?"

"Nope, I don't. Do you?"

"Yep. Cuz he doesn't wanna be spotted!"

Claire steps out of the van, skips all the way down the sidewalk, and turns back to smile and wave before disappearing into the school.

I grin at the joke, and grimace at having put my head in the bear trap again. I think of my mother. She has never micromanaged or second-guessed my parenting, but she once shared an excellent piece of advice: *don't make promises lightly, and keep the ones you make.*

I do keep the ones I make, but I make more than I should and have relied on sheer luck to bail me out too many times. There's a lesson here, waiting to be learned.

It's smooth sailing all the way to the airport. I circle up the ramp and find a spot on the north side of Level Four in the short-term parking garage. From here it's a direct walk across the northern skybridge to an escalator that drops you a few steps from Delta's check-in counter.

My tools are checked, and I breeze right through security. My scopes don't warrant a second look this time. Today I'm the windshield, not the bug, and I hope it stays this way.

At the gate, I listen to a voicemail from Cortney Kinman. The vault in Albuquerque didn't open this morning, confirming that the problem is not a timelock overwind. We are green-lit for next week; I'll book a flight for ABQ as soon as I can.

As boarding begins, I receive an email from the lawyer in Minneapolis, double-checking our Sunday appointment. I'm tempted to prank her, but decide against it. "We're good to go," I reply. "See you at the airport on Sunday."

I'm waitlisted for an upgrade, but it doesn't happen. I'm stuck in 29E, a middle seat toward the back of the plane, between two dudes that are as overweight as I am, maybe a little more. Lovely. Consoling myself with the irrelevant truth that we will all land at the same time, I put my cell in airplane mode, cue up a playlist, and lean back in the headrest as we taxi across the tarmac.

Eva Cassidy pulls me away from my pity party over first-world discomforts. It doesn't matter how many times I hear her sing "Autumn Leaves," it always knocks me out. I had heard many renditions of this song, and played it myself, both instrumentally and behind vocalists. But I'd never *felt* it until I heard Eva's version, with its breathy, beautiful tones, especially in the coda.

Had Blue Note Records signed Eva when they had the chance, she might have been the first multi-genre vocalist to have hit records in every major field. She sang everything, from jazz to R&B to folk to country to rock. When she resisted being pigeonholed into a single category, though, Blue Note declined. Eva died of cancer a few years later, shortly after this song was recorded, live at Blues Alley. She was thirty-three. Tragically, all of her commercial success has been posthumous.

Time for a mood change. Almost any comedy will suffice. Laughter, even of the muted kind required by airplane etiquette, keeps the mind light and limber, which can be helpful on a day with a potentially problematic lockout.

But dang, I'm exhausted. This has been a grueling week, and Eva's voice is so soothing. I close my eyes for just a minute . . . and wake up more than an hour later to the captain's voice, urging passengers to view some topographical wonder out the right side of the plane. Eva's playlist is over.

I hate falling asleep in the middle seat. More accurately, I hate *waking up* in the middle seat, and having that moment of panic about possibly invading someone's space. The window is best. You can lean against it on a folded blanket or jacket and conk out without a worry in the world.

In the middle, you can't move in either direction. Head check: I'm contained in my own seat, not drooping over or resting on a stranger's shoulder.

Leg and arm check: legs are under the seat in front of me, and arms are crossed, hands still clasped together, resting on my stomach.

Mouth check: no drooling. This is good.

The world comes into focus. There's no time for a movie now, but that's okay. We'll be landing soon, and I still need to do my final job prep. Problem is there isn't much to prep. The vault door won't open, and the question is why. I'm hoping it's a sluggish timelock. Robert Coleman has already massaged the door with a sledgehammer in an attempt to shock-activate the timelock. He and I will do it again just to be sure.

If it works, Robert and I will be in and out quickly, with time to catch up over a cup of coffee or a coke. I could conceivably be at the airport five or six hours before my flight. That would be fantabulous.

If the sledgehammer fails, we'll have to drill. I'm not sure where, but we aren't going to have a replay of Vegas — no possible way are we running short on drill bits today. I brought at least a hundred, mostly short ones but also some footlongs and a few eighteen-inchers, all in several diameters. Can't imagine needing the long ones on this job, but I'm not taking any lazy chances. We're *not* running dry.

Hope for the best, but prepare for the worst.

For the first time since getting the call, it occurs to me: more than any safe or vault I have cracked in recent years, this one has extreme and opposite potentials. It could conceivably be snap-easy with a few sledge-hammer blows, or it could turn into the worst nightmare of my career. There's also the worrying possibility of it handing me my first bank vault failure.

Better think positively.

CHAPTER TWENTY-ONE

Robert and the Full Automatic

THE PLANE SUDDENLY LURCHES HARD LEFT, THEN DOWN, INTERRUPTING my Zen time. Welcome to the skies above Salt Lake City, where your stomach can jump into your throat at any second.

Descent can be bumpy, but the aesthetic trade-off is worth it. I glance around the neighboring beer belly, out the window to white-capped mountains in every direction. It's breathtakingly beautiful, and it's easy to see why Brigham Young was smitten enough to settle here. Everyone should land at SLC during daylight at least once.

I pull out my phone and turn off airplane mode the second we hit the runway. Amazingly, there's only one text. It's from my oldest daughter, Jenny, asking if Claire can stay over tomorrow night. Claire will be thrilled, but I need to find out from Val if there are any scheduling conflicts. Before I even notice that Jenny's text was to both of us, Val responds affirmatively. Awesome.

Jenny spent four years attending medical school in Israel, and only recently moved back to Portland, where she is finishing her pediatric residency at Oregon Health & Science University. Claire and her brothers have enjoyed getting to know their older sister again, and a sleepover is a special treat.

Robert is looking his usual professorial self when I spot him waiting for me at baggage claim. Put this gray-maned, elegant man in a tweed jacket with elbow patches, and he could walk into the faculty room at any university without anyone raising an eyebrow. I told him so years back when we first met.

"How was your flight?" he asks as we shake hands.

"Dunno, I slept through it," I reply, smiling.

As we approach the carousel, I see both bags. Wow, that was fast. What a relief, especially after Monday's scare when my Beemer case went temporarily AWOL. I lift off the DeWalt tool bag containing all the drill bits, set it down, and reach for the handle on the Beemer case. I spot that red ribbon, frayed and stained, well past its prime. Dang, I keep forgetting to replace it.

Robert's van is an exact copy of Hitch's. White Ford with a blue Diebold logo on the front doors. We pull into traffic, and I realize that I don't have a clue how long the drive is. "Hey Robert, how far are we from the jobsite?"

"From here maybe six or seven miles. The bank is right downtown."

"Oh, that's great."

"About this bizarre lockout, Dave. What's the plan? Do you have a drill point in mind?"

"Not yet," I reply. "Figured we'd put hands on it together and come up with something onsite."

"Wow, that's not your SOP, is it? I mean, all the vaults we have opened together, I'm pretty sure you walked up to each one with a drill point ready to go. Am I wrong about that?"

"You're right. But this is an unusual case. It's difficult to figure out what the problem is. With no dials or handle to use for diagnostic purposes, all we can do is make some educated guesses."

"You mean we're gonna wing it?"

"Yup."

Robert looks over. "This'll be interesting."

"Let's just hope it's not *too* interesting. Hey Robert, I'm curious about something."

He looks over again. "Yeah? What's that?"

"Well, you've lived here for a long time, haven't you?"

"Sure have," he says.

"Just curious. Are you LDS?"

Robert smiles. "Sure am. Why do you ask? You too?"

"I was once."

He glances over. "Once? As a youngster?"

I nod.

"What happened?"

"It just didn't ring true. But maybe I'm too picky because *no* religion has tenets that ring true to me."

Robert chuckles. "Tough sell, are ya?"

"I suppose so."

The van slows, and Robert pulls into a parking stall. We're onsite in beautiful downtown Salt Lake City. At the back of the van, Robert hands down the Beemer case, my carry-on, the DeWalt tool bag, and my laptop case. I unzip the carry-on, lift my briefcase off the spare change of clothes, and put the suitcase back in the van. Won't need the clothes unless I spend the night, and that's not happening. I have to get home tonight — I promised.

A bank employee recognizes Robert the moment we come through the door. She rushes over to usher us downstairs. And there it is. The majestic Diebold full automatic vault door.

My goodness, it's way taller in person than it looked in Robert's photograph. I grab a tape measure out of the Beemer case.

"Told you it was a big 'un," says Robert, smiling.

"You were right," I reply. "This beast is nearly eight feet tall."

I have been in this business for a long time. It takes a real challenge to get my adrenaline pumping. Well, it's pumping — a little *too* much. I need to touch, play, and be active for a bit, to make the distracting tingles subside.

I reach out, grab the captain's wheel, and rotate it clockwise as hard as I can to seat the door as deep in the jamb as possible. Then I try rotating counterclockwise. It doesn't budge — this vault is solidly locked. Time to treat it to some Richter-scale vibrations.

I pick up the long four-by-four Robert leaned against the wall yesterday when he was done. On it I draw two concentric circles with a black Sharpie marker, and a solid ball in the center. A bulls-eye. "You ready?"

"Sure," Robert replies. The goal is to shock-activate something. We don't know what, exactly. Maybe the timelock. Maybe the boltwork-actuating device. *Something.*

The Diebold full automatic in downtown Salt Lake City. It has no dials and no handles. Just a captain's wheel for easing the door in and out of the jamb during opening and closing.
Source: Author photo.

I stand next to the giant crane hinge, grab one end of the four-by-four, and hold it horizontally across the middle of the vault door. The bulls-eye is directly above the captain's wheel. I nod to Robert. "Let 'er rip!"

He picks up the sledgehammer, pulls it back like a baseball bat, and lets loose with a slo-mo Barry Bonds swing that nails the center of the bulls-eye. The crack is deafeningly loud. Our escort darts back up the stairs. Robert grins. I would, too, but my fingers are stinging.

"Holy cow!" I exclaim, laying down the wood and shaking my hands. "Paul Bunyan's got nothing on you, Robert!"

I hold the wood steady and steel myself for imminent pain. Wish I had gloves. Robert unleashes a dozen blows so brutal that the four-by-four starts to split at the edges. I lay down the wood and rub my numb hands together.

I approach the captain's wheel. If we got lucky during the sledge attack and the boltwork retracted, the wheel will rotate and the door will come out of the jamb.

No luck. The captain's wheel stops immediately. Robert comes over and places his electronic stethoscope at several locations on the door. He shakes his head. "Nothing, Dave. This timelock isn't ticking."

"Okay then," I reply. "We'll assume that time ran down just like it was supposed to and that the problem, whatever it is, is not traceable to a sluggish timelock."

"Yeah, that's gotta be right."

"Robert, I'm going to take an hour or so to commune with this old girl, try and get in the flow of her energy field, and see if we can make friends." I bite the inside of my lower lip to maintain a straight face.

"*What?*" Robert furrows his brow.

"I'm going to spend some time working up a drill point."

He nods and smiles. "Good one. Ya got me."

"Don't think I have a spiritual side, eh?"

"Well, *you* might, but she, er, *it*, doesn't," he says, gesturing to the vault.

Robert reads something on his phone and looks up. "Dave, I have another bank downtown that just escalated a work order to top priority. How about if I run over and take care of their problem and bring back a late lunch? By then you'll have a drill point, and we can get to it. Does that work for you?"

"Sounds like a plan, man. And food-wise, Robert, I like everything. Whatever you're getting, just double the order."

"You got it," he says, heading for the stairs. "See you in a bit."

I pull up a chair in front of our obstinate opponent and pull out my laptop. This is only my second drill job on a full automatic bank vault, but I have documentation on several dozen.

I mentally review how a full automatic works. On the other side of this foot-thick door is a timelock riding atop an automatic boltwork actuator. The gap between them is a little less than half an inch. The trigger arm protrudes from the top of the boltwork actuator into this gap, where it sits next to the timelock's release lever.

When time runs down, the release lever moves sideways and pushes the trigger arm over. This activates the first of many small levers inside the bolt actuator. The levers move together in a coordinated ballet, ultimately firing the powerful boltwork retraction spring, which retracts each of the two dozen round door bolts to unlock the vault.

That's how it's supposed to work. Something has obviously gone awry. But *what?*

CHAPTER TWENTY-TWO

Skip's Maxim

ONE OF MY OLDEST FRIENDS IN THE BUSINESS WAS THE SUPREMELY talented Skip Eckert, a kind and generous soul who passed from this world entirely too soon. Skip won our industry's annual manipulation contest every single time he entered it. Like boxing's Rocky Marciano and Floyd Mayweather, he retired undefeated, and the list of technicians he vanquished in competition reads like a veritable Who's Who of elite safecrackers. Skip was also the first professional to drill open a bank vault and free a trapped child.[1] We chuckled over that one for years.

In the late 1980s, we placed a half page ad in a locksmith magazine. Using a map of the United States, split in half at the Mississippi River, we put Skip's photo and phone number on the right side and mine on the left. The pitch was to locksmiths nationwide. If they needed help on a difficult lockout, Skip and I were available to board a plane on a moment's notice. It worked wonders for both of us.

Skip and I were chatting one day about the difficulty in determining causation when a certain set of symptoms present themselves on a particular model of vault. He made a pithy observation that became an instant classic between us. I call it Skip's Maxim: *When in doubt, suspect the timelock.*

Skip's Maxim relies on the fact that timelock problems cause more bank vault lockouts than any other single cause. Playing the odds is a useful strategy in both gambling and safecracking.

Skip's Maxim is our best bet right now, so we are going to drill for the timelock. What and where is our target? We are aiming for that half-inch gap between the timelock and boltwork actuator. The hope is to get close enough to the release lever and trigger arm to move either one over and fire the actuator.

There is just one problem: I don't know the height of that gap. And even if I did, I don't have the left-right measurement either. It's frustrating. On nearly all vault doors, the locations of critical components are standardized within tolerable variations. Full automatics are the exception. No two in my database are exactly the same.

If ever a bank vault were a riddle wrapped in a mystery, inside an enigma, it'd be the full automatic with mysterious malfunction. The key to solving it is accurate information. And maybe a dash of luck.

I had two dashes with me a few years ago in Wallace, Idaho. I was at a Wells Fargo in Wallace's historic district, where *everything* is an antique. It's like walking onto an Old West movie set, but it's all real.

To confirm that the lockout on their Diebold full automatic wasn't due to an overwind, Wells Fargo waited five days, the maximum that can be wound on that particular timelock, plus one extra day just to be sure. The vault didn't open. The service company had already put sledgehammer to door. No joy.

I needed but didn't have precise information on the location of the timelock or boltwork actuator. What to do? What else — guess and go for it!

The captain's wheel assembly is attached to the door with four half-inch diameter screws. I removed one and drilled down its threaded hole with a smaller bit so as not to bugger the threads. It took footlong bits and quite some time to penetrate the entire door. When I looked in with a scope, I could tell that the hole was too low and about two inches away from the trigger arm. Except for a small sliver of air at the top, all I could see looking straight in the hole was the bronze case of the boltwork actuator.

I lassoed the trigger arm through that tiny gap with a small loop of music wire and pulled. But it didn't move — it was already in the unlocked position. The timelock release lever had moved it over as intended, but, for some reason, the automatic boltwork actuator didn't retract the boltwork.

The full automatic bank vault in historic Wallace, Idaho, where the entire downtown area is on the National Register of Historic Places.
Source: Author photo.

The smell of defeat was in the air. It looked like my perfect record might come to an end right there in Wallace. But I wasn't about to call a coring company without first exhausting my bag of tricks. So I threw a Hail Mary.

In the center of this automatic vault is a timelock and a boltwork actuator that retracts and extends each of the twenty-four bolts around the periphery of the door.
Source: Author photo.

I turned my drill bit around and ran the butt end into the hole until it touched the bronze case of the boltwork actuator. I then turned my hammer drill on full blast with gentle forward pressure. After a few seconds, I heard the rumbling of the boltwork trying to move, and then it stopped.

I tried to pull my drill bit out of the hole, but it was stuck. *Really* stuck. I sprayed Tri-Flow down the flutes, locked ViseGrips on the long shaft, and twisted the carbide bit back and forth until it finally came out. The rumbling started again as the boltwork slowly retracted.

I reached over, rotated the captain's wheel, and swung open the vault door. By this time, it was late in the evening, so it was just the branch manager and me. She waltzed over and gave me a huge hug. "Right now," she said, "you're my hero."

In truth, I felt more lucky than heroic. But as far as Wells Fargo is concerned, I knew what I was doing all along. That there was any luck at all involved, well, let's keep that our little secret. I kind of like being thought of as a knight in shining armor.

The problem was easy to spot upon disassembly. The timelock had worked flawlessly. Its release lever had moved sideways and pushed the trigger arm just like it was supposed to. But one of the tiny levers inside the boltwork actuator had cracked and bent, effectively stopping the ballet mid-performance, until it was nudged along by vibrations from my hammer drill.

The local company took the damaged component to a machine shop. A brand-new one was made the next day, and this beautiful 1890s full automatic bank vault was back in operation, ready to provide another hundred years of service.

My streak remained intact.

CHAPTER TWENTY-THREE

Plan A and Plan B

THE PROBLEM IN WALLACE ENDED UP BEING THE BOLTWORK ACTUATOR rather than the timelock. But that was a one-in-a-million fluke. Here in Salt Lake City I'm playing the odds, and the odds favor a timelock problem. Its release lever probably didn't push the trigger arm over to start the chain reaction inside the actuator.

Plan A is set, then. Drill for the trigger arm that sticks up into that half-inch gap between the boltwork actuator and timelock. I don't know its precise location, but it's somewhere in an area about one-foot square, right in the middle of the door. That's a start, but I need to narrow it down. There are more than fourteen *hundred* places to drill a 5/16″ hole in an area that size. That's far too many. I don't want to Swiss-cheese this beautiful antique.

I will occasionally use video footage from a bank's security system to examine the back side of a vault door. But the camera is on the wrong wall to give me the view I need. Dang, I'd give anything for a peek at the other side of this monster.

As if on some divine cue, my vest pocket begins to vibrate. I usually let calls go to voicemail when I'm working. But for some reason, I answer.

It's Hank Taggart, Diebold's account rep for this bank. "Hey Dave," he says, "just checking in to see how things are going."

"Hi, Hank. Well, it's going slowly. Our attempt at a quick opening failed. I'm trying to work up a drill point now, but it's tricky because I don't know the precise location of the primary locking components inside this door."

"Would a photograph help?"

My heart skips a beat. "What? You have a photo?"

"I do," Hank replies. "I was there a few weeks ago and fell in love with that gorgeous old vault. The craftsmanship in the boltwork is something to behold, and I snapped a photo while I was there. Would you like me to text it to you?"

"Does a bear poop in the woods? Hank, you may have just saved the day."

Pure gold appears on my phone's screen a few seconds later — a photo of the boltwork, complete with Hank's smiling reflection off the see-through glass panel on the back of the door. Back in the day, vault manufacturers were so proud of their doors that they made the back panel out of glass so bankers and patrons alike could view the impressive bolt-work. No manufacturer does this today. You can probably guess why.

Hank's photo is incredible, truly one of the greatest gifts I have ever received in the middle of a tough job. There is no need now for educated inferences or wild-ass guesses. I can see everything.

I'm transferring the photo to my computer when Robert trots down the stairs carrying food. I didn't realize how hungry I was until the aroma hit my nose. It's time for a break.

Robert pulls up a chair next to mine and glances at the photo on my laptop screen. "Do you think that's similar to what we have?"

"Better than that, Robert. That *is* what we have."

"What do you mean? Did you find a match in your digital library?"

I explain how our incredible stroke of luck came to be, and Robert is as gobsmacked as I was. He distributes deli sandwiches and mustard, and we gaze at a single photograph we wouldn't trade for all the coffee shops in Seattle.

Like the Wallace vault, this one's boltwork actuator sits very close to the horizontal centerline of the door. That makes sense — the springs in the actuator have to push and pull twenty-four heavy round door bolts in and out. The closer to center, the more equally distributed is the stress on the actuator.

The difference is that the actuator in this door is not on the verti-cal centerline; it's about six inches to one side. Without the photo from

Hank, I would likely have drilled the same hole here that I did in Wallace, and it would have yielded a view of absolutely nothing, well behind all the critical components. Safecrackers call that an exploratory hole, or complimentary ventilation. Anything but a mistake. We are a prideful bunch.

Of interest to me is the Burton-Harris boltwork actuator. This gives us a way to accurately date the vault door. The Burton-Harris was introduced in 1887; Diebold began making and using their own actuator in approximately 1891. We can thus conclude that this full automatic was manufactured sometime in that four-year window — a trivial but cool fact.

With a digital ruler on Hank's golden photo, we calculate a drill point for the trigger arm. When the sandwiches are gone, I look over at Robert and raise my eyebrows.

"I'm ready if you are," he says with a determined grin.

"Let's do it."

I carefully measure and Sharpie-mark the drill point and hand the tape measure to Robert. While he works to confirm the location of our drill point dot, I pull out my phone and open WWF. Whoa! Valerie has chosen to skip her turn and draw seven new letters. She never does that! I play ZA (up) and ZOA (across) for forty-three points and the lead. Yes!

Robert tells me our dot is perfect. I triple-check, and we are ready to rumble. Although this vault door is nearly a foot thick, it's a little less than eight inches through the many layers of high-grade steel to our target. The gap between the timelock and boltwork actuator is about half an inch, so accuracy is mandatory. If I angle up or down at all, we'll miss the gap.

I fire up the Fein motor and make a crater right through the dot using a 3/16″ bit. This provides a starting point for the 3/8″ bit I swap in. Take a breath, begin a slow exhale, and pull the trigger. We are instantly rewarded with bits of metal falling onto the shop rags below. The first layer isn't super-duper hard, which makes for a great start.

Robert spots me from the side. His job is to make sure I don't angle up or down as the carbide drill bit penetrates deeper and deeper into the door. When his thumb points down, I lower the back of the motor until he gives me a flat palm. When his thumb points up, I raise the motor until he gives me a flat palm. Straight is the grail.

The next layer is much harder, so we switch to a new, razor-sharp carbide bit and kick the Fein into high speed. It's slow going at first, but small chips and powder come gushing out the hole when the pressure gets high enough.

We go back and forth, through lamination after lamination of super steel, alternating speeds and swapping out bits. We finally break through ninety minutes later with an eight-inch bit pulled out from the chuck as far as it can go. I instantly back off the lever bar, but it's too late. The bit binds, and the motor twists in my hands for a few seconds before stopping. It's stuck.

Stuck isn't a bad thing — it usually means the bit is intact rather than broken, which is a fantastically *good* thing. Another half second, though, and it would have snapped off. I remove the keyless chuck from the bit, lock ViseGrips on it, and twist back and forth to work the bit out of the hole. It's badly bent, but still in one piece.

I open my scope case to grab an arthroscope and feel a mild rush. The tingles are back. This natural high is downright addictive and undoubtedly explains my longevity in the industry.

As the scope slides down the hole, I see the various steel laminations in the door. They look like tree rings. Then I spy what caused the bit to bind. We came in a fraction of an inch low, so the very bottom of the bit snagged the very top of the boltwork actuator's bronze case.

We are pretty much where we want to be, in that small gap between the timelock and boltwork actuator. It's a beautiful sight. The trigger arm is just off to the right of the hole. I expected it to be a little further to the right, but this is fine.

Next step is to lasso it. I take an eighteen-inch length of music wire, bend an inch-long "V" on the end, and put a gradual arc in the last three inches. The "V" shape enables the wire to be shoved in a small hole, pointy end first, and the arc causes it to drift over toward the trigger arm once it pops through inside the door.

I watch through the scope as the V-wire lassoes the trigger arm on the very first try. I tug on the wire, but it feels solid — the trigger arm isn't moving. I tug harder but get the same result, so I lock a pair of ViseGrips onto the wire and pull hard. Too hard, actually — the "V" straightens out and the entire wire comes out of the hole. *Hmmm.*

I glance over at Robert, who notices a concerned look on my face. "What's wrong?"

I exhale loudly. "The trigger release isn't moving."

"Why is that?"

I shrug my shoulders. "I'm not sure, unless it's already in the unlocked position."

That's what happened in Wallace.

"Is there a way to check?"

I nod.

I open the scope case and swap the arthroscope for a longer, retroview cystoscope. I don't use it often, but the retro view can be a lifesaver when you need to look back toward yourself. I slide it slowly into the hole, past the timelock and boltwork actuator. I follow the path of the trigger arm, and, sure enough, the timelock release lever is right next to it — quite clearly in the *fully unlocked position.*

"Well, mister," I begin, with a grimace. "Everything we've done thus far was for naught. The timelock did its job and time ran down. The release lever pushed over the trigger arm, which should have fired the boltwork actuator and retracted the bolts. Robert, this vault should be open."

He reaches over and tries the captain's wheel, but it doesn't move. "What's Plan B, boss?"

I have a Wallace-inspired idea. "There is something we can try," I reply, looking at the pile of used drill bits on the floor. I grab a 5/16″ bit and chuck it up, but backward this time, with the butt end sticking out. I insert it into the hole until it contacts the bronze case of the boltwork actuator.

"Let's try a little vibration." I put the Fein in hammer-drill mode and pull the trigger. The noise is deafening as the mini jackhammer unleashes its fury on the vault. The boltwork actuator is definitely feeling these shockwaves. *Come on.* This trick worked in Wallace, and I need it to work here, too. But after a full minute, nothing has happened. I stop and sit to think. And rest the ears.

"What are the possibilities, Dave? What else could it be?"

I glance at my watch. It's five-thirty. Wow. What happened to the day?

"Robert, it's possible something in the boltwork actuator broke, not allowing it to fire. And it might be the case that vibration alone won't overcome it." It crosses my mind: *if that's the problem, we may have to bring in a coring crew.*

"It's also possible that a screw backed out somewhere in the door. It could have dropped down into just the right place to block the boltwork from moving."

Robert makes a sour face. "It could be anywhere, right? Oh man, that would be difficult to overcome."

I nod. "Yeah, especially on a door that takes so long to penetrate. It could take many, many holes to find the blockage and remove it." It crosses my mind again: *if that's the problem, we may have to bring in a coring crew.*

These negative possibilities are making my butt pucker. Privately, I have to acknowledge the elephant in the room — my undefeated bank vault streak looks highly vulnerable right now. I much prefer being in control, knowing what the problem is and how to solve it.

"Here's another possibility, Robert. Maybe twenty-three of the twenty-four door bolts *are* retracted. A sole problem child may have disconnected from the rest, and it could be holding the door behind the jamb. I have had orphan bolt disconnects on bank vaults around the country, but never on a full automatic. There's always a first time, though."

"Is it different on a full automatic?"

"Sort of. On many vault doors, we can shim between the door and frame with thin plastic or paper. That allows us to feel any bolts that are extending behind the frame. But antique vault doors use a pressure system to suck the door tight against the frame. There isn't enough of a gap between the door and its frame to slide a shim through."

"They made 'em tight, huh?"

I nod. "This vault was manufactured during the golden age of nitroglycerin attacks, back when tolerances had to be extremely tight to prevent soup from being poured in the cracks."

Robert mulls this. "Okay, then how would we go about finding a disconnected bolt?"

"We'd drill a few exploratory holes and carefully scope the door. Eventually, we'd spot the culprit and then drill another hole or two, to retract the orphan bolt."

He shakes his head. "Let's hope that's not it. Otherwise, we'll be here all night and into tomorrow."

I think of Claire, but opt for brevity. "Robert, my flight is at ten this evening, and I really want to be on it. Let's get this door open."

Robert drums his fingers along the top of the captain's wheel. "Alrighty then. What's our best Plan C?"

I rub my temples. "I don't know yet, but we aren't giving up without a fight."

Chapter Twenty-Four

Plan C: The Clock Is Ticking

I THINK BACK TO SAFES AND VAULTS FROM MY PAST WHERE THE automatic boltwork actuator malfunctioned. I can remember only one, and it wasn't a bank vault. It was a century-old cannonball safe in the County Treasurer's office over in Stevenson, about an hour from my home.

Time ran down, and the timelock triggered the automatic boltwork actuator. But the two large, round door bolts in that old safe failed to retract: the post that holds the powerful retraction spring had fractured and the spring slipped off.

I drilled a hole through the side of the cannonball and used an extra-long pin punch to tap on the hinge-side door bolt. I wasn't sure if this would also retract the bolt on the other side, but it did. That was a sweet opening.

Perhaps something similar happened here. Maybe the post rounded off or broke. Or maybe the spring itself broke. Either of those things could cause all twenty-four door bolts to stay locked behind the frame.

Could the technique I used on that old cannonball safe be adapted to this situation? Not really. I can't get to the side of the vault door to drill directly for one of the bolts. There is no *side* to speak off — just a long wall on either side of the door.

I could angle-drill through the jamb, but only with exact measurements for one of the door bolts. Even with precise measurements, though, it would be risky. We'd have to hit a bolt in just the right spot for a long punch to drive it back without slipping off, given the steep angle the hole would be at. The target is just too small.

I glance at my phone. It's almost six o'clock. I open the lid on my laptop and bring up Hank's photo again. I notice the eight bolts along the opening side of the door. They're all connected to a long, thick vertical strap that runs nearly the full height of the door and looks to be about three inches wide.

"Need your opinion, Robert." An idea is beginning to form. "I'm thinking of starting a hole near the edge of the door and angling toward the middle. I wouldn't be aiming for a door bolt but rather for this strap." I point to the long vertical strap in the photo. "If the angle is steep enough, we might be able to punch back the strap and carry all the bolts back at the same time. What do you think?"

"Not sure, Dave. Um, isn't that strap connected only to the eight bolts on the opening side of the door? What about the bolts on the hinge side, and the ones along the top and bottom? Do you think all twenty-four will retract if we punch on that one strap?"

"I'm not sure, but it's worth a try. What do you think?"

"I dunno. You'd be drilling on a forty-five-degree angle, which adds almost fifty percent to the drilling depth. Do you have any footlong drill bits with you?"

I dig through the Beemer case. "Exactly eight of them. Four 5/16″ and four 3/8″."

Robert glances at the pile of ruined bits it took to drill a mere eight inches straight through the door. "I hope that's enough."

"Me too. The good news is that we won't need those long bits until the short ones bottom out."

I remove a large sheet of blank paper from my laptop case and begin plotting the drilling angle. We know the exact depth to the strap, and also its left-right location, since we can see the strap through the scope, next to the hole we drilled for the trigger arm. That hole didn't open the vault, but it's now providing us with crucial information for Plan C.

I draw a forty-five-degree line from the crudely drawn strap until it intersects a second line representing the front of the door. According to this top-view drawing, we need to start the hole 2 5/8″ in from the edge of the door. Following this template should put us right in the middle of the strap.

Using a fat black Sharpie, I simplify our drawing to an easy-to-see isosceles right triangle. The hypotenuse represents the angle we need to maintain throughout the drilling process. I measure over 2 5/8″ from the edge of the door at about chest height. Robert tapes the triangle to the door, and I commence penetration. There's no time to waste.

This is a comfortable height — no stooping over or balancing on the tips of my toes like a ballerina. We cruise right through the first lamination. As expected, the hardened next layer is much more difficult to penetrate, especially at this forty-five-degree angle.

I swap out a five-inch bit for an eight-incher and glance at the time. It's nearly seven o'clock. I gently lean into my lever rig, and the tip of this brand-new bit snaps off. Must have been defective out of the box.

A short blast from a can of compressed air sends pieces of the broken tip flying out of the hole. I chuck up another eight-incher, carefully insert it into the hole, and hit the trigger on my Fein. As soon as I apply pressure to the drill motor, this bit grabs and breaks just like its predecessor. *What's going on?* Again, I blow out the pieces, grab my scope to look at the bottom of the hole, and spot the problem: we are half on the edge of a hardened, horizontal plate.

Great. Drilling hardened material at an angle is difficult enough. But being on the *edge* of such a plate, and at such a steep angle, is worse. Much worse.

"Whatcha thinking?" asks Robert. "Should we start over with a new hole?"

What I'm thinking is that I'd like to borrow the Oerlikon cannon from Clint Eastwood in *Thunderbolt and Lightfoot* and blow a hole in the wall. Or the thermal lance from James Caan in *Thief* and cut the stupid door in half.[1] Or some water bombs from Robert De Niro in *The Score* and reduce this entire vault to rubble.[2]

But I say something different. "It took us more than an hour to get this far, Robert. Starting over would send us back to the drawing board to deduce a new drill point and a different angle." *Which would mean missing my flight.*

"Let's stay the course."

Robert sticks a bit in the hole and compares its angle to the triangle drawing. "Yeah, we're right on the money. We just need to get past the corner of that plate."

Kicking myself for not bringing any diamond bits or carbide burrs, I chuck up another eight-incher and switch the motor into high speed. I hope the extra RPMs will cause the bit to penetrate at least a little before dying. It does. And so does the next one and the next one and the next one. And the one after that.

I feel a double vibration in my vest pocket. Doesn't matter who it is or what they want, it'll have to wait. We continue drilling. And breaking bits. But we are making progress, slowly but surely.

I glance at my phone as we resume drilling. It's two minutes shy of eight o'clock. On my lock screen, I can see the text from Val's phone: *Hi Daddy. This is Claire. I was just wondering, how is your job going?*

It's going badly, and I don't dare reply. Claire may be harboring a misconception about how easy Dad's job is. She traveled with me a few months back to a far-off job via Amtrak, the one and only time I've been unable to book a flight. We had a sleeper car, and it was one of my favorite work trips ever. The opening went well — she even helped me drill a small hole in the vault. We were in and out in less than two hours.

Last year, Claire and I traveled to a family Thanksgiving in Kansas City. (Val and the boys had flown in the day before.) We landed in the late afternoon, and I listened to a panicked voicemail from Bank of Nevada. They had closed for the day, wound the timelock, and shut the vault — only to hear the shouts of a teller trapped inside. Was there anything I could do?

Luckily, the teller had her phone inside the vault. I called, and she described the door, which enabled me to identify make and model. I guided her to the release button to unlock the timelock. Employees on the other side of the vault were then able to dial the combinations and turn the handle to open the door. The entire episode took about five minutes, and the teller was free to spend the Thanksgiving holiday with her family.

Claire probably has it in her head that I snap my fingers and bank vaults fall open. There are rare days when it may seem that way, but today

Claire shows good form as she helps me drill a bank vault in Whitefish, Montana.
Source: Author photo.

is definitely not one of them. Robert and I aren't even halfway through drilling this stupid godforsaken hole.

That's it — *no more Mr. Nice Guy*. I motion Robert over. "We're deep enough now that I couldn't change the angle if I wanted to. No need to spot me. But I sure could use you here."

I point to my lever rig. "Let's put Paul Bunyan on the end of the bar."

Robert smiles, walks over, and grabs it with both hands.

I chuck up one of the eight-inch bits I used earlier, and he leans into the bar. Like Hitch in Vegas, Robert in Salt Lake understands the importance of applying even pressure, gradually increasing it as needed. Sudden or jerky movements are a no-no.

Penetration goes smoothly. We swap out bit after bit and change drilling speeds over and over again. When an eight-incher bottoms out, I replace it with a twelve-incher and stair step the diameter from 3/8″ down to 5/16″, ensuring sufficient chip relief as we head into deep territory.

We proceed through lamination after lamination until a twelve-inch bit finally bottoms out, and I have a moment of pure god-awful panic. *There are a few eighteen-inch bits in the Beemer case, right?*

Yes, there are four. Two of them are 3/8″ in diameter and and two are 5/16″. The deepest part of the hole is only 5/16″ in diameter now, so the 3/8″ bits won't fit. Which means we have two usable eighteen-inch bits. Two. *Seriously?*

Another double vibration in my vest pocket, which I ignore. I dare not waste another second. I chuck up that big boy, ease it to the bottom of the hole, and hit the trigger. It goes in about a quarter-inch before the sound of the motor changes. We are breaking through. And then the tip snaps off. *This is not happening!*

Safes and vaults are inanimate objects made mostly of steel and utterly devoid of conscious intentions. But I swear, sometimes there's an evil genius inside, plotting a devious counter to my every move.

A look in the scope confirms my worst fears. The entire carbide tip has basically welded or wedged itself in the very bottom of the hole. *Dammit!*

I try my best to maintain my composure, but I'm so pissed off that I don't even try to speak. I have broken more bits today than I have in the past year. This is ridiculous. Then again, so is drilling through laminations of hardened steel at a forty-five-degree angle.

I grab one of my super-long, homemade 1/4″ pin punches and lock ViseGrips on the butt end. I insert the punch down the hole until it contacts the broken piece of carbide and give it one hard whack with my four-pound hammer. I had intended on giving it more, but it's not necessary. The punch plunges down the hole and drives the piece of carbide into the air gap beyond. The tip of the drill bit must have broken off just as it was breaking through to the inside, and the punch drove it through the remaining eggshell.

We just got majorly lucky. Again.

I chuck up the last eighteen-inch bit in 5/16″ diameter, flip the switch to high speed, and clean out the hole. Holding my breath, I insert a cystoscope. It slides through into the air gap and I see perfection personified. We have hit the very center of the three-inch-wide strap controlling all eight door bolts on the opening side of the door.

"Robert," I say, "the hole came out perfectly. We are *exactly* where we want to be."

He nods, and I pull out my phone. It's twenty after nine — time's up. And there's another text: *It's me again. I know you are working Daddy, you don't have to answer back. I just wanted to say goodnight and I love you! See you in the morning!*

I want to reply, but we're out of time and then some.

I hand the extra-long homemade punch to Robert, along with my four-pound hammer. "I'm going to start packing up. Put the punch all the way in the hole. It'll seat into a nice crater that we just drilled into the strap. Do your Paul Bunyan thing one last time. I mean, beat it like you mean it. If we catch a break, you'll hear a rumbling sound and the boltwork will retract."

I turn to pack. This job started out so organized, with cases and tools nicely arranged on the floor. It's now utter chaos, with drill bits, scopes, punches and miscellaneous crap scattered everywhere. Putting it all back in the Beemer case is going to take a few minutes.

Oh my. I'm used to loud noises, but Robert's rapid-fire, metal-on-metal banging sounds are excruciating. I pivot my battery doors outward to lessen the impact.

This is it. The moment of truth. If the vault opens, I have a chance of making my flight — scheduled to leave in thirty-two minutes. If it doesn't open, I'll have to stay over, break my promise, and work on an impossibly stubborn vault tomorrow, with a scary high risk of failure. *Come on, man, be optimistic — it's gonna open.*

I'm almost finished packing when Robert taps me on the shoulder. He probably called my name and I didn't hear him. I quickly push my battery doors in and turn to face him.

"Dave," he begins, frowning. "The punch is stuck."

I was afraid of this. The quarter-inch punch was too thin to survive his Paul Bunyan blows, and it probably bent way down deep in the hole. *Or maybe not.* Maybe the boltwork is trying to retract but is being stopped by the punch, à la Wallace?

I rush over and lock a pair of ViseGrips on the end of the punch. "Robert, was there a rumbling sound when the punch got stuck?" In Wallace, the boltwork made all kinds of noise as the bolt actuator took over and the twenty-four large, round door bolts struggled to retract.

"I don't think so. I didn't hear anything at all."

We twist the punch back and forth a few times to get it out, and I cock an ear toward the vault. No rumble. Nothing but stone-cold silence. The boltwork isn't even *trying* to retract.

My shoulders droop as I glance over at Robert. "Well hell, Pilgrim, it looks like we're coming back in the morning."

My neck and ears are on fire as I turn back to finish packing. Need to call Val and Claire to break the news. And get a hotel. There's no hurry now, and no point in sugarcoating it. Tomorrow is gonna suck. I dread having to work on this #$@&%*! vault again. What the hell is holding it locked? I wish I knew.

I stand the Beemer case upright and slide my laptop case over the handle. I need some cool air and a warm bath. And maybe an edible. I turn around, temporarily defeated and out of gas, but doing my best to not let it show. *Never let the client see you sweat.*

I look up, and there stands Robert, posing next to a wide-open vault, sporting an even wider smile. "I thought you were *never* going to look over!"

I stand there frozen, trying to process this impossible scene. *There was no rumble.* In a flash, it occurs to me why: in Wallace, the long rumble was the sound of a boltwork sluggishly retracting as the actuator limped along with a bent internal component. But here in Salt Lake City, the actuator isn't limping along. It can't do *anything*, because its retraction spring is either disconnected or broken.

Robert hammered this massive boltwork all the way back with no help at all from the actuator. The punch got stuck when it was driven too deep, *after* the twenty-four door bolts were fully retracted. He had the good sense to try the captain's wheel — it spun.

Like the full automatic in Wallace, this one in Salt Lake City has twenty-four bolts that lock the door behind the jamb, controlled by the timelock and the boltwork actuator that reside in the center of the door.
Source: Author photo.

The vault is open.

The deeper meaning washes over me like a wave. *Pinky swear!* I leap into the air, throwing a fist-pump like we just won the Super Bowl. Robert and I lock hands and bring it in for a brief and modern man hug.

My elation turns to alarm. "Robert, what time is it?"

He looks at his watch and sucks air through his teeth. "We have twenty-eight minutes, Dave."

All I can do is hope for the best. We grab the cases and hustle up the stairs. And for the first time in my life, I'm running through a bank like a robber bolting for the door. Only faster.

I hope it's not too late.

SATURDAY

CHAPTER TWENTY-FIVE

Treat Time

"Daddy, are we going to Dairy Queen?"

I meet her gaze in the rearview mirror.

"I don't know, sweetie. Do you *want* to go to Dairy Queen?"

"Is Grandma there?"

"Not on Saturday."

"Okay. Can we go through the drive-thru, then, and get a treat?"

"We can do that," I reply, smiling. "I know what *you* want!"

Claire and I are just leaving the dance recital. It was adorable. Few things in this life are as precious as a stage full of little girls twirling and clicking and doing their best to curtsy in sync. I snagged a center seat in the front row, and for once, the video I shot doesn't have adult heads bobbing along the bottom of the screen.

We pull up to the menu board. "Welcome to Dairy Queen," says a young woman through the intercom. "I can take your order whenever you're ready."

"Okay," I reply, pausing to choose between single flavor and swirl. "We'll have one small vanilla cone and a butterscotch Dilly Bar, please." Claire *loves* butterscotch.

"I saw you videoing, Daddy. Can I watch it now?"

Not the best timing, I think, as we stop at the window. But I'm a softie. "Of course, sweetie," I reply, and hand her my phone. "But be careful not to get ice cream on it, okay?"

"I'll be careful. I promise."

Yes, I made it home. The last person on the plane, I wasn't the slightest bit bothered that my upgrade was given to another passenger. They were on time to the gate; I was not. It mattered not a whit that I was in 32B, a middle seat in the last row. I was just happy to be in the air, en route to Portland, in time for the recital and with my perfect record on bank vaults still intact.

I got lucky. The plane was delayed a few crucial minutes — I wouldn't have made the flight otherwise. And it's a good thing a cop wasn't glancing in the bank's window when Robert and I were sprinting across the lobby, carrying bags and wheeling a suitcase. Suspicious behavior? Sheesh!

Truly, though, from the very beginning of my career, I've been blessed with crazy good luck. I've snatched victory from the jaws of defeat so many times that I almost expect it when my back is against the wall. Like it was last night.

Close as it was, though, it wasn't the *closest* call I've had. Flight attendants didn't shut the cabin door behind me for at least twenty or thirty seconds. A few years back, however, I had a bona fide buzzer beater — a vault that opened with *no* time left on the clock.

True story. But before I tell it, I should confess: there are no safes or vaults to crack today. Claire and I are going to finish our treats and join Val and the boys for the second half of their soccer game. When it's over, we'll run Claire home to change, then hustle to *her* game. And when that's over, she'll go spend the night with her older sister.

At dinner, Val and I will pop the cork on a silky red blend, then kick back to play board games with the boys. Although we're benevolent authoritarians about most things, we're quite democratic when it comes to choosing games. Everyone gets an equal vote. I'm hoping for Catan or Codenames, but the majority will probably overrule me and choose Monopoly.

All of this brings me to say: I'm going dark for most of today. In the meantime, I'd like to share a few of my favorite short and supershort stories — about jackpots, penetration parties, a bomb that didn't explode, incredible kindness, off-the-charts luck, the closest call of them all, and more.

Some of these disclosures are a little embarrassing, but I'm old enough now to shrug off a few lapses in judgment from my past — even the doozies that cause me to shake my head in disbelief. We'll reconvene late this evening to make final preparations for tomorrow's lockout in Minnesota, which promises to be a unique and memorable challenge.

Chapter Twenty-Six

Jackpot!

Safes with unknown contents are usually filled with sailboat fuel and a paperclip. But once in a while, it's KA-CHING! I have seen safes jam-packed with cash, drugs, coin collections, jewels and gems, stock certificates and bearer bonds, rare baseball cards — including Mickey Mantle's rookie card in mint condition or better, worth beaucoup bucks — and more.

My favorite such sighting occurred in Reno, Nevada, where a young couple hauled two beefy safes home after winning the high bid at a Bay area auction.

They picked me up at the airport, and I did my best to prepare them for disappointment. But my talk of scant odds did nothing to curb their enthusiasm. They had heard things rattling around inside one of the safes during the move and just *knew* it was the motherlode.

We pulled in the driveway, and they ushered me into the garage, where I went to work on safe number one — a century-old Herring-Hall-Marvin beauty that manipulated in about twenty minutes. Hubby excitedly swung open the cast-iron door to reveal stale emptiness and nothing else, not even a paperclip. This wasn't the safe from which those tantalizing sounds had emanated, so my hosts shrugged off the dud and transferred their optimism to our next opponent.

Made in Finland, this modern, high-security container boasted a German-made keylock, an American-made combination lock, a glass plate, and multiple relockers. I opened the combo lock first, and spent the next two hours trying to pick the keylock before giving up and drilling a scope hole through the side of the safe into the stubborn lock. With eyes now inside, I made a minor adjustment to the pick, it turned, and the lock opened.

"We're just about there." I said, "I'll get the safe fully unlocked, then you can pull the door open."

"Awesome!" said Wifey. "Um, can I take a photo first?"

"Of course," I replied. Hubby sat down next to the safe and peeked through the scope sticking out the drilled hole, while I swiveled on my chair to face the camera.

The second commandment of safe and vault work: **Remove scope from hole *before* pulling the door open.** This is to prevent damage to a brittle and costly tool.

As you may have guessed, I broke that unforgiving commandment. After the camera clicked, I turned back to the safe. I planned to turn the handle and swing the door out just far enough to ensure it was fully unlocked and openable, but without exposing its contents.

The door didn't budge so I tugged harder and soon realized what was stopping it. I let go of the handle, but it was too late. My decapitated scope was already on its way to tool heaven. I had a brief out-of-body experience myself, because that particular scope — a brand-new Storz Swing Prism — was the single most expensive piece of equipment in my entire safecracking arsenal.

Despite my best effort to not let it show, my gracious hosts detected despair in the air. And they empathized. Indeed, they forgot all about the treasure hunt until I violated my own protocol and grabbed the handle to pull the door all the way open. I intended on retrieving the other end of my broken scope and getting out of their way. But when the door opened, diamonds and tigers and pearls — oh my! — spilled out all over the floor. Well, not tigers.

From the paperwork inside the safe, we learned that it had belonged to a jewelry store in San Francisco back in the 1980s. For reasons unknown, it ended up in storage. Many years later, trailing a long-delinquent storage bill, the safe was sold at auction.

I repaired the drilled hole and serviced the safe, while Wifey began adding up the retail value of the merchandise. (The tags were still on

Ten seconds after this photo was taken, I turned around and pulled open the door — and broke the scope that's protruding out the right side of the safe. The scope had been inserted through the edge of the door and into the keylock. When the door was pulled open, the scope snapped in two like a candy cane. Ugh!
Source: Author photo.

many of the pieces.) When Hubby and I departed for the airport, she had crossed into six figures with no end in sight.

What makes this job so memorable isn't the mini-motherlode (I've seen many), but the attitude and maturity of this young couple. They didn't talk animatedly about spending their newfound fortune on lavish vacations, fancy cars, or flashy bling. Material possessions weren't even on the radar. No, they were over the moon simply because college for their two young daughters was now secure.

For that reason, this is my all-time favorite jackpot story. Well, that and maybe also the fact that my generous hosts tipped me nearly enough to replace that broken scope.

Every so often I'm asked, *What's the most money you've ever seen in a safe or vault?* Well, excluding the Federal Reserve Banks — whose vault doors we serviced when I was with Allied Safe & Vault — where bills are stacked in carts and on pallets in dollar amounts I can scarcely fathom, and excluding the big bank currency centers with their enormous piles of cash, the answer is seven million dollars.

It was during a drug bust at a large mansion in the affluent Eastmoreland neighborhood in Portland, Oregon. I rolled up and was met by special agents from both the Drug Enforcement Agency (DEA) and the Internal Revenue Service (IRS). When I asked a guy in an IRS jacket why they were there, he shrugged his shoulders and said, "Taxes." Even drug dealers are supposed to report their earnings. This particular supplier was obviously high up the food chain, raking in huge money, as evidenced by a multipurpose recreation room that was nearly the size of my entire house.

The safe had gone undetected until a sharp-eyed agent noticed that the wall in the rec room didn't extend quite as far as the wall in the hallway outside it did. He grabbed a tape measure and confirmed his suspicion — the hallway wall was several feet deeper. He soon discovered why: one of the bookshelves along the rec room wall was attached to a cleverly disguised door on hinges. It pivoted to reveal a long, narrow room containing one of the tallest burglary-resistive safes we manufactured at Allied.

The agents were stacked behind me, watching over my shoulder as I drilled on the hellacious hardplate known as Maxalloy. Once penetration was complete, I slid my arthroscope into the hole, aligned all the tumblers, and opened the lock. But before backing away to let the agent in charge pull the door open, I turned in his direction and said, "When your boss called, she mentioned that I had my choice of charging the usual flat-rate fee or opting instead for ten percent of the contents. In my shoes, what would *you* do?"

Flummoxed, he struggled to process it until he noticed the smirk on my face and waved me away. We traded places, and he swung the door to reveal a large interior that was packed top to bottom with nothing but money and drugs. Mostly money. I didn't find out the total until I watched the evening news. That bust netted seven million dollars. In cash.

Oh yeah, ten percent would have suited me just fine.

CHAPTER TWENTY-SEVEN

Penetration Parties

I'M NOT A FANCY DRESSER, SO I AVOID FANCY FUNCTIONS. BUT IN 1988 AN industry friend persuaded me into attending the annual black-tie shindig hosted by the Greater Philadelphia Locksmiths Association. In a rented, heavily starched tuxedo, I found myself seated at a large round table next to Ed Dornisch, owner of Keystone Safe Company. Over the years, Ed had collected eighty locked safes, and he asked me for a price to come back to Philly and open them all. I gave him a number, and we struck a deal before the second course arrived.

Also seated at the table was Rex Parmalee, a long-time industry instructor who frequently taught classes on manipulation and safe lock servicing. He overheard the conversation and shared an idea. "What would you guys think about turning this into a hands-on class?" he asked. "You provide the drill points, and students penetrate the safes. It could make for a helluva fun party." Ed and I liked the idea, so Rex got on the phone the following week and promptly sold out the class.

A month later, forty safecrackers from around the country convened in Ed's warehouse, ready for action. The very first safe we attacked at the very first Penetration Party was a tall Mosler whose door we pulled open to reveal the most extensive, privately owned coin collection most of us had ever seen. There were stacks and stacks and more stacks of completed coin folders, and hundreds of rare coins in plastic holders. An amateur collector in attendance guesstimated the value at half a million but said it was probably higher, since there were many coins about which he had no knowledge.

Ed had purchased this safe at an estate sale. The Mosler and its contents were legally his. But because he was that rare human being whose moral compass always pointed north and never wavered, not even in the face of extreme temptation, Ed called the family. "You guys might want to come down and see what your father left behind."

You were a good man, Ed Dornisch, may you rest in peace.

In three decades of combining classroom lecture with hands-on training, I've seen only one injury at a Penetration Party. That's a surprising statistic, given that we are a bunch of adrenaline junkies wielding power tools, having the most fun a person can have with their clothes on. I suppose part of the explanation for why there haven't been more injuries is that we talk openly and at some length about safety, about maintaining control of one's emotions when a job goes south, and about the importance of taking a break if that control starts to slip. With only one exception, it has worked well.

About that exception: a rookie making a rookie mistake? Yes to the rookie mistake, no to the rookie. It was me and, try as I might, I can't find anyone to blame but yours truly.

It happened during a Penetration Party at Empire Safe's warehouse in the Bronx, where I was demonstrating a technique for opening one particular model of ATM. The idea is to slice through a thin piece of rubber next to the dial and peel it back far enough to drill a hole and bypass the locking mechanism. The hole is then repaired, the rubber is rolled back in place, and no visible evidence of the breach remains. Properly executed, it's a good trick. Improperly, well, not so good.

I held the box cutter in my right hand, with the tip of the brand-new razor blade resting on the rubber. I then laid the back of my left hand on the rubber, about two inches below the box cutter, turned to the class, and said, "Remember: always cut away from, and never toward yourself." But rather than demonstrating what to do, I proceeded to show the class precisely what *not* to do. I sliced downward, executing a perfect cut through the rubber, continuing on into the index finger of my left hand, starting at the tip and exiting at the second knuckle. Penetration to the bone.

You could have heard a pin drop in that room full of safe technicians. I'm sure I initially turned red from embarrassment, but my face soon went white when the guy who doesn't like the sight of blood found himself bathed in it. That huge gash *gushed*.

What to do? Fortunately, we had just moved from the classroom to the warehouse, so the guys grabbed their tools and started working on safes. I excused myself to the nearest sink and held my hand under the faucet for a few minutes to clean the wound. I had full range of motion and minimal discomfort, so I was confident there was no tendon or ligament damage. Shane Ellison, one of my dearest friends in the industry, butterflied the wound shut in several places, waited for the bleeding to stop long enough to apply a coat of superglue, and then gauze-wrapped the entire finger.

I went out and finished the demonstration on the ATM, which, to my relief, went flawlessly. After swinging the door open, I smiled at the group and said: "I know you all know this, but I'm going to say it anyway. When you have one of these buggers to open, please please please, *do as I said, not as I did!*"

While motherlodes are rare at Penetration Parties, safes with unknown contents are not. The problem is that every locksmith and safe technician knows from experience that most such safes are empty. Which is why, every now and then, I like to spice things up and have a little fun.

"Ladies and gentlemen," I will announce excitedly, "I have fantastic news. Last week our host bought several safes at an estate sale. One of them is rumored to have somewhere between two and three hundred grand inside. For real."

And then comes the motivational carrot. "Our host is not telling us which safe is supposed to be loaded, but he *is* making us an unbelievably generous offer. Get a load of this, gang: he's offering a finder's fee to the lucky safecracker. Yes, ten percent of the take goes to the tech who opens the jackpot safe."

Ten percent of two hundred grand is twenty large. Drill shavings are soon flying in every direction as students frantically drill for dollars. Eventually it's all grins and groans when the lucky safecracker swings open the door to reveal two (or sometimes three) hundred grand.

Jackpot!
Source: Author photo.

CHAPTER TWENTY-EIGHT

Three Species of Luck

I COULD WRITE AN ENTIRE BOOK ON NOTHING BUT THE FANTASTIC, freakish luck I have enjoyed in my career. Maybe I will someday, but, for now, here are three examples of what I'm talking about, each one representing a different species of luck.

First up is luck in the sense of extremely low probability — getting a vault open with a million-to-one shot. I was at a bank in Lake Oswego, Oregon, with Perrill Smith, the Diebold tech I have worked with most frequently over the decades.

Both combination locks were functioning properly, but the vault door would not open. Perrill and I got down to business and went through the usual suspects in the usual order, ranked by probability. First, following Skip's Maxim — *when in doubt, suspect the timelock* — we drilled for the timelock. It was unlocked. Then we drilled for the two spring-loaded relocking bolts in this particular door. Neither was fired. Then we drilled for a device called the "daylock." But it too was working fine.

At this point, we had been onsite for six hours and drilled four holes in the door. We had made zero progress — the door was locked up just as tightly as it was when we walked into the bank. I was out of good ideas, the bank was running low on patience, and Perrill really wanted a burger.

I decided to drill an exploratory hole over near the opening edge of the door. To modern, professional safecrackers, an "exploratory hole" has one purpose: it provides an insertion point for a variety of scopes with different angles to give us different views of a safe or vault's locking mechanism. I wanted to look up and down the door, close to the jamb, to check for any abnormalities or apparent problems in the door's bolts. Under normal circumstances, this hole would be placed close to the midline of the door to yield equally good views up and down. But my back was hurting, and I didn't want to stoop over any more than absolutely necessary. So I made a Sharpie dot about chest height on the door, at a spot I figured would be suitable for scopes. It was more of a guesstimation than a drill point.

I attached my portable drill press to the door, fired up the Fein, and commenced drilling hole number five. The 5/16″ specialty bit went through the outer steel skin, then nearly a foot of high-PSI concrete, and finally through the inner steel skin. Once it broke through, I released the trigger on the Fein and grabbed one of the spokes on my drill press. I intended to rotate the spoke and back out the drill bit so I could put a scope in the hole for a look around. But instead of the drill bit coming out, *the vault door pulled open!*

I looked over at Perrill. He cocked an eyebrow, looked at me, then at the open vault, and then back at me. We locked gazes and broke out in spontaneous, hysterical laughter. Mr. Corey would have frowned at our acting like toddlers for a few brief seconds, but we couldn't help it. Pretending that *we meant to do that* was impossible.

The question Perrill and I had was obvious: how did an exploratory hole open the vault? I mean, it hadn't yet been used for exploring! The answer was evident once we swung the door all the way open and looked at the door and its jamb. The door's bolts were in the *unlocked* position the entire time. The door jamb was two pieces, and one of them, a long piece of stainless steel that runs full height, had come loose at the top and sprung out askew behind the upper two door bolts.

On many vaults, this wouldn't have caused a lockout. But on this one it did. You see, with this model, even when the door's bolts are retracted, they still stick out a little bit. Not enough to hang up on a standard door jamb, but enough for that tall piece of stainless steel to spring over and lock behind them. To this day, I can hardly believe it happened.

Had I drilled an eighth of an inch to the left, the bit would not have pushed the stainless piece over far enough to clear the door's bolts; an eighth of an inch to the right and the bit would have gone through the stainless instead of pushing it over. Had the hole been any lower, it would have missed entirely; any higher and the stainless would have been pushed even more behind the door bolts. The drill bit had to bump the stainless *exactly* where it did.

What made this opening so bizarre is the fact that I did nothing else but drill the hole. I didn't insert a scope or turn the dial. I didn't push or probe anything. Heck, I didn't even turn the handle! This is the only safe or vault I have ever opened simply by drilling a hole, and I seriously doubt there will ever be another.

Perrill got his cheeseburger (with extra pickles). I had a BLT and onion rings. We both wanted a beer but settled for ice water. And we cracked up with nearly every bite. Even now, years later, we still shake our heads at the insane luck that propelled us to victory that day.

Second up is odd-duck luck — making a bone-headed mistake and having it work far better than *not* having made the mistake. I was with Jim Wicken from Cook Security, and together we confronted an old Diebold vault suffering a timelock malfunction.

Before I go any further, know this: I'm not perfect. I commit the occasional blunder, as is well attested by examples throughout this book. But one mistake I just don't make is drilling at a mismeasured drill point on a difficult-to-penetrate door. The reason I don't make that sort of mistake is that I check, double-check, and triple-check the location of my Sharpie dot. Although I don't bother doing this on mid- and low-range containers, I do it on every high-end safe and vault. If the drill point is four inches left of dial center, then, by gawd, my Sharpie dot will be measured out and marked four inches left of dial center. If I make a mistake, I'll spot it on the second or third pass and correct it. This happens all the time.

On this particular day, though, I learned that triple-checking doesn't help if I fail to spot a directional error at the outset — *the drill point said* **right** *but I went* **left***!* I made a Sharpie dot on a piece of blue painter's tape, just like I always do. Then I double-checked and triple-checked the location of that dot, failing each time to notice that it was in the wrong direction. The likely reason I didn't spot the error is that I had drilled the same make/model vault across the country a few days earlier, and it was hinged on the opposite side. On that door, left was correct.

In any event, I didn't notice the mistake until the hole was about ninety percent complete. Then it hit me like a brick. I was majorly perturbed, and almost started a second hole at the correct drill point. But for some strange reason, I didn't stop. I finished drilling and stuck a scope in just to confirm the bad news.

But when I looked in the hole, everything was wrong. The model of timelock was wrong. The blocker and snubber bar were on the wrong side — everything was turned around backward. *What the heck?* It turns out, the door had been retrofitted with the new timelock reversed from normal, and guess what: my "mistake" had miraculously put me *exactly* where I needed to be!

Mr. Corey would have been proud — I kept a straight face, reached in with one of my homemade tools, and unlocked the timelock. Jim pulled open the door, looked at the retrofitted timelock, then over at me. I just grinned and pretended omniscience. And floated all the way home on Cloud Nine.

My third and final example is luck in the sense of a buzzer beater — getting the vault open just as time expires. I have beaten the clock many times, but this was the closest call of them all.

Bank of America was locked out of a vault four hours north of me and needed it opened immediately. The rush was for a panicked groom who needed access into his safe deposit box to retrieve a wedding ring — he was getting hitched the very next day. Wonder of wonders, BofA had already green-lit the job, so I headed north immediately but hit a traffic jam behind a wreck not far from my home. I called to let the Diebold tech and the bank know I would be arriving a little later than anticipated.

I pulled into the parking lot at four-thirty. Dann Kain, a former Seattle cop turned bank service technician, met me at the door with bad news. The bank had gotten nervous after my delay and called a coring crew for insurance. The crew was already onsite with clear instructions: if Dave doesn't have the vault open by six-thirty, take over the job site and core a gigantic hole through the wall. *Here we go again.*

I was not a happy camper. Had I been informed of these conditions while en route, I might have turned around and gone home. But I was at the bank with all of my tools, within spitting distance of a challenging vault that was tickling my amygdala. Leaving would have felt like a big fat fail. So I stayed. And the clock started ticking.

While the coring crew was examining building blueprints to determine which wall to breach and where, I was reviewing photos and drawings to plan an attack through the door. I had no idea what was causing the lockout, and it was one of the PITA vaults with not one but *two* glass plates and a couple of spring-loaded relockers. I could open it by breaking one or both glass plates and dealing with the relockers, but this would require far more time than had been allotted.

I saw one shot at success. If I could drill a hole in between the two glass plates, in just the right spot, I might be able to pry up the blocker bar in the vault and then turn the handle. The blocker is a piece of one-inch square bar, about eight inches long. When it is lifted high enough (by the combination locks in a functioning vault), the handle can be turned to retract the boltwork.

The blocker is a nice big target, but the hole would have to be perfectly placed for me to pry it up far enough. I scaled off my photos with a digital ruler, transferred the measurements to the door, and made a Sharpie dot at the drill point. Because of the ticking clock, I didn't even bother double-checking (much less triple-checking) myself; I started drilling immediately.

Fortunately, bank vaults with glass aren't nightmares to penetrate, and my footlong drill bit pierced through the door a couple of minutes after six o'clock. I glanced over at the coring crew. They were done strategizing and were now lugging equipment into the branch. They planned to breach the wall very close to the vault door, which meant I wouldn't be able to continue working once they began. Wiggle had left the room. The six-thirty deadline was now in stone.

I looked in the hole with a cystoscope. It was good news/bad news. The good news was that the hole had hit the bulls-eye — we were right on the blocker. The bad news was that there was a piece of stranded cable in between us and the blocker. I hadn't been careful enough when the drill bit penetrated the inside skin, and it had nicked the cable. Two of the seven strands of stainless steel were severed. One end of this cable is attached to one of the glass plates; the other end is connected to a spring-loaded relocker. The cable had enough intact strands to hold together, but I didn't dare nick it again. (The other relocker is attached to the second glass plate, whose cable I had safely missed.)

I carefully inserted a long, thin screwdriver, nudged it past the cable and under the blocker. I twisted the screwdriver to raise the blocker, but it would lift only about halfway. I needed a way to exert greater upward force on it. The hole was 3/8" in diameter, and so I used my hand grinder to put an angled ramp on the tip of a long 3/8" pin punch. I inserted it into the hole and gently wiggled it past the cable, with the tip of the ramp under the blocker. As I tapped the punch in, its ramp was lifting the blocker. When the punch was all the way in, I triumphantly grabbed the vault's handle, but it wouldn't turn. The blocker needed to be raised another sixteenth of an inch or so to create enough clearance for the handle to turn.

It was now six-twenty, and I was beginning to feel a little anxious. If time hadn't been an issue, I would have enlarged the hole from 3/8" to 1/2", being extra careful not to sever the remaining strands of the wire that was holding the relocker in check. The larger hole would allow the insertion of a larger diameter pin punch, which in turn would raise the blocker higher and permit the handle to turn. But with ten minutes left on the clock, this wasn't an option.

I had one move and one move only. I motioned Dann over and put him on the handle. I was going to hammer in the ramped punch, hoping that the sudden, upward momentum would bounce the blocker up that extra fraction of an inch we needed. I wasn't sure the blocker *could* be bounced, but it was the only thing I could think of on the spot.

I inserted the ramped punch into the drilled hole, wiggled it past the cable, and placed its knife-edge tip under the blocking bar. I then ran a Sharpie marker all the way around the punch, right at the edge of the hole. I smacked it hard with my four-pound mini-sledge. The punch went in an inch, and Dann tried the handle. No go. I pulled the punch out just far enough to expose the Sharpie ring, and we repeated this again and again and again, with Dann mixing up the timing on the handle each time. No go.

I haven't worn a watch since getting a smartphone, which was in my pocket, so I asked Dann. He tilted his wrist. "It's twenty-six after," he said, ruefully. I looked over, and the coring crew had finished their preparations. They were ready to take over.

I was super frustrated. There was no doubt in my mind I could open the vault. I just needed an hour to enlarge the hole while carefully avoiding the cable, and then raise the blocker that little bit more so the handle could turn. But the bank wasn't going to wait. In four minutes, they were moving to Plan B.

There was nothing left for us to do but keep doing what we were doing. As time ticked away, I felt panic adrenaline run out from body center to the extremities and cause an involuntary shiver. I began rapid-fire punching, pulling out, punching, pulling out, while Dann was frenetically jerking the handle back and forth. I glanced over and spotted the branch manager and the foreman heading our way. Time was up. We were being shown the door.

Mad as hell, with my jaw so tightly clenched that my teeth hurt, I pulled back the mini-sledge and let loose with one last KAPOWZER-BAMMO! Dann immediately stumbled to one knee. It took me a second to realize what had happened: he had jerked hard on the handle, but instead of encountering resistance like he had previously, the handle turned all the way unlocked, and the momentum caused him to lose his balance.

"It's open!" he shouted, bracing his weight against the door. Stunned but elated, I rolled up the shop rags with drill shavings trapped inside and stood back as Dann slowly walked that big door nearly one hundred eighty degrees until it butted up against the heavy-duty doorstop against the wall.

Had the vault not opened, I would have gone home with nothing to show for it other than a broken streak. But it *did* open. The bank was pleased as punch, and their safe deposit client was over the moon. He thanked Dann and me profusely on his way into the vault and again on his way out, holding the little ring box up to cheers from us all.

The closest call of them all. This vault door opened at the very last possible second, and I felt like Tom Brady after yet another miraculous comeback.
Source: Author photo.

Hail meet Mary. That's the way my career has gone. The goddess of safecracking has always watched over me, time and again preventing me from teetering into disaster. I feel like a miracle man — charmed, blessed, and grateful.

CHAPTER TWENTY-NINE

BOMB! and Miscellany

IN THE EARLY 1990S, A FRIEND AND FELLOW SAFECRACKER HAD AN eye-opening experience that motivated many of us to change our door-opening protocol. Carl Cloud spent his career in and around San Diego, opening many safes for local law enforcement. All were routine except for one old antique safe on a drug bust. Carl swung open the main door and encountered a full-height inner door secured with a keylock.

Law enforcement is always anxious and a little on edge during a bust, so we try to get them in as quickly as possible. On this particular keylock, the fastest way to open it is to punch it. Carl had a pin punch in one hand and a hammer in the other, but he paused when it occurred to him that the perp was in the back seat of a squad car. Carl suggested that the officers check his keyring. They did and found the key.

When Carl pulled the inner door open, he noticed a suspicious-looking device attached to the back of the keylock. Like any experienced safecracker, Carl knew what teargas looked like, and this wasn't teargas. So the officers called in the bomb squad. And sure enough — it was a *bomb*, and it almost certainly would have exploded had Carl punched the keylock. He was shaken for weeks and shared this harrowing experience with his colleagues.

Many of us decided to forego brute force attacks in situations where a booby trap is a concern, such as when a safe is being opened under warrant. Some of us went even further and adopted a global *'somebody else pulls the door open'* policy, just in case. Self-preservation is a powerful motivator.

That's the deal behind the reveal.

A DISCOMFITING SPOTLIGHT

My most uncomfortable moment on a job wasn't when a SWAT team surrounded the bank after a duress alarm accidentally triggered while I was drilling inside the vault, or when I used a thermal lance on a supermarket safe and burned up a pile of money, or even when I dropped a heavy safe and chopped off half my finger. (That would have been curl-the-toes distressing, but the nerve was in the severed finger under the safe, not in the remaining stump, so I felt nothing.)

As uncomfortable as those situations were, none of them caused me to wake up in a cold sweat months later. This one did. It was during a drug bust at a biker bar in southeast Portland. I walked in with my tools and encountered a large group of law enforcement officers wearing ski masks, and four perps seated on couches along the wall.

I was in my early twenties but looked fifteen. When they saw me and realized why I was there, the perps began mean-mugging and swearing up a storm. *That effing punk kid ain't opening the mother-effing safe!* I was taken aback by the hostility and couldn't tell if they were upset that I was going to open the safe or skeptical that I would succeed.

I called over the commander and asked, "Hey, what's with the masks?" He leaned close and said, "We're taking measures to protect our identities." I was speechless. It felt like an episode of *The Twilight Zone* or *Candid Camera.*

"So, next month or next year," I said, regaining my voice, "it'll be *my* face they recognize in a restaurant or supermarket? Seriously? You guys could have given me a heads-up. I own a darn ski mask."

I felt like walking out. But I stayed and finished the job and was a little jumpy in public for a while.

Over the decades, I've opened many safes for law enforcement. To my knowledge, I've never bumped into any of the many people who owned the safes. But I'd like to think that, even if I did, they'd understand — I'm just a guy doing a job.

KINDNESS AND KARMA

After a job in Philadelphia, I was in the air, scheduled for a short layover in Chicago before catching my flight home. As we approached O'Hare airport, the city was obscured by a wall of white. The snow was thicker and heavier than I have ever seen, including many flights to jobs in Alaska.

Ours was the last aircraft to touch down before the airport closed to traffic. No planes were taking off or landing. There were zillions of people in a dozen lines at United's customer service center. Some of them were not very well behaved, and United's agents endured a ridiculous amount of abuse for something that was beyond their control.

Hours later, when I got toward the front of the line I was standing in, I couldn't help but notice how calm and empathetic United agent Larry Haynes remained, despite dealing with passengers who were shouting and slapping the counter and acting like entitled asshats. Larry's imperturbable kindness impressed me, and I conveyed the sentiment when it was my turn. He thanked me and apologized for some bad news — at present, the earliest he could get me home was the day after tomorrow. But he did tell me to check back in with him about five in the morning, as he had heard a rumor that United might be opening up another flight to Portland.

No fan of crowds, especially grumpy ones, I went in search of two empty, adjacent seats. I found them at the very end of Concourse C, in the waiting area for Gate 31. I watched *Meet the Parents* yet again, set my phone alarm for five o'clock, and dozed off.

I awoke to the words, "Mr. McOmie, Mr. McOmie." It wasn't a dream, it was Larry. "I'll trade you," he said. I sat up and shook out the cobwebs. "I'll trade you this ticket for the one I gave you a few hours ago," he said, smiling.

Turns out, United *did* open a new flight to Portland. Larry must have remembered my name or written it down, because he put me on that flight, and in first class to boot. And then, rather than wait for me to come to him, he took his lunch break to find me.

I was blown away, genuinely touched by this kind gesture. Six hours later, I was in the friendly skies of United, heading for home. I recounted this story in a customer compliment to United, and sincerely hope Larry got a raise, or at least a nice company perk. Karma is on his side.

A Life Lesson for a Young Safecracker

Although this isn't a book about politics or economics, I'd like to share an experience I had as a young safecracker. It had an immediate and long-lasting effect on my thinking. Your experience and viewpoint may differ, and that's just fine — the world would be a boring place if we all thought alike.

The story begins with my lifelong friend Rick Wise. After our wild, lock-picking year at WWU, Rick, savvy as ever, finagled his way into a subsidized home loan from the Farmer's Home Administration (FmHA). His mortgage payment was just over a hundred bucks a month. For real.

Envious, I applied and got one of my own, but there was a catch. The maximum you could earn in a year was a little under twenty thousand dollars. By this time, I had left Mr. Corey and was working as an eager beaver young safecracker at Allied Safe & Vault. With an hourly wage of seven dollars and ten cents per hour, I was well under the line.

At the end of the fiscal year, though, the company looked at my productivity — I was fast and efficient — and gave me a whopping forty percent raise to an even ten bucks per hour. I did the math and discovered to my consternation that this juicy raise put me just over the limit. I would lose that precious subsidy.

So I walked into the boss's office and asked for a fifty-cent *decrease* per hour. Max Kirklin, the vice president of operations and my direct superior, cocked his head as if he hadn't heard me correctly. "Let me get this straight," he said. "You want a *cut* in pay?"

"Yes sir, I do."

Max just stared at me for a few seconds. He reminded me of Noah Bain from *It Takes a Thief* — balding, selectively gregarious, and ever so slightly imperious. The perfect boss, really.

"I've been managing this branch for a very long time," he said, smiling. "But this is a first. No one has ever come into my office and asked for *less* money."

I smiled back, meekly.

"I'm happy to take off the fifty cents, Dave," he said. "But, do you mind telling me *why* you want to do this?"

Max listened patiently, drumming his fingers on the arms of his leather high-back chair. I explained the FmHA loan subsidy, the income limit, and how nine-fifty an hour would keep me below it. When I finished, he nodded slowly and said, "I see."

Max paused for a few seconds, parked his elbows on the desk, clasped his hands together, and rested his chin on his fingers. "I hope you see the long-term problem with this strategy," he said, his eyes unwavering. "*Do* you?"

The puzzled look on my face telegraphed my answer. "I'm not sure."

"First of all, Dave, remember this: government money comes from taxpayers like us. Whenever something is subsidized, it just means taxpayers are paying the difference."

It was obvious once he said it, but I hadn't thought of it that way before. I nodded in agreement.

"That isn't the question I want you to grapple with, though," he said. "You're a young man with your entire career in front of you. What about next year? And the year after that? You're very good at opening safes and vaults. Will you be happy *never* getting another raise?"

I just sat there, dumbfounded and a little embarrassed. But I was too prideful to withdraw the request. Max honored it, and I kept my subsidy, at least for a while.

A few weeks later, I told Rick what Max had said. "He's right," Rick replied. "Almost everyone around me totally builds their life around the subsidy."

I noticed the same thing. Most of my neighbors were on the same FmHA home plan, and they were absolutely paranoid about losing it. They took seasonal jobs or worked for cash to avoid going over the limit.

On one of my many trips to the local FmHA office, I asked an administrator about the program. "The original idea," she said, "was to give young homeowners a chance to get into a house with no cash up front. They could build a little equity, and use their portion of it as a down payment on their next house with a regular mortgage. Hopefully, in five years or less."

I couldn't resist. "How often does that happen?"

She paused before answering. "Almost never," she said, ruefully.

I have no doubt that the program was started with the best of intentions, but what a disaster. Rick and I both bailed soon after, opening up spots for others to wear the shackles of self-imprisonment disguised as self-improvement. I couldn't see it any other way — I felt I'd been duped.[1]

To one degree or another, we are all products of our past, and this part of mine played a pivotal role in shaping my fiscal conservatism. I was lucky to have a boss as plainspoken and diplomatic as Max Kirklin. Suddenly awakened to the power of incentives, disincentives, and paths of least resistance, I went from being oblivious to seeing the obvious.

Funny how that works.

EXHAUSTION AND MENTAL ACUITY

The lockout was in Onaway, Michigan. My original flight was canceled, and I ended up on a red-eye from Portland to Chicago. Unfortunately, my sleeping plans were foiled by a newborn in the row behind me who screamed the entire way. Poor thing must have been in excruciating pain.

I arrived at O'Hare exhausted, waited three hours for my connecting flight, and then made an hour-long drive to the bank with Gary, a technician from Security Corporation and my assistant for the day.

The vault contained a glass plate that we had to break to reach our target. Attached to the glass plate were four large spring-loaded relockers that fire when the glass breaks, and each one of them must be neutralized before the vault can be opened.

But I had something up my sleeve. I noticed long ago that, if I drill at high speed through tempered glass with a hot carbide bit, the glass often stays intact for a minute or two before shattering. When I'm on my game, I can drill through the glass, continue on to my target, and turn the vault's handle *before* the glass shatters and the relockers fire. You see, once the handle is rotated and the boltwork is retracted, it doesn't matter if they fire — the vault is already unlocked.

Everything was going according to Plan A. We drilled through the vault door and the glass plate and came out directly on target. I inserted a long pin punch and tapped the obstruction out of our way while the glass remained intact. And then I suffered probably the biggest brain fart of my career, which is saying a lot, because there have been many. In my sleepless daze, *I forgot to turn the handle!* I stood there for at least three full seconds before I realized that the last step had not been taken. I lunged for the handle just as the glass plate exploded, firing the relockers. A split second made all the difference.

Plan B necessitated drilling for those four dastardly doohickeys, which in turn required several cans of Rockstar, multiple pots of coffee, and an equal number of trips to the restroom. Though not the longest in terms of duration, this was the most draining job of my career. I need my beauty sleep!

SAFECRACKER ≠ LOCKSMITH

As a teenager, I was a decent locksmith. I could install deadbolts, pick locks, open and make keys to American-made cars, and perform reasonably competent security surveys. Today, no shop would hire me based on my locksmithing ability. I dabble on occasion but can barely pick the easiest of locks, and I haven't kept up with more than four decades of sweeping changes in the industry.

Some locksmiths are safecrackers, but most are not; some safecrackers are locksmiths, but many are not. The terms simply aren't synonymous. Twice I have locked myself out of my work vehicle. On both occasions, I called locksmith friends to let me in, just like they call me to open tough safes and vaults.

The first time it happened, I had just pulled into the driveway after flying home from a difficult vault job back east. My neighbor, Joe, who was principal of the middle school at the time, walked over to greet me. I stepped out of the van to say hello, and for some dumb reason, I hit the LOCK button and shut the door — with the engine still running.

"Oh crap," I said. "I need to call one of my buddies to come let me in. They'll get a chuckle out of this."

Joe was confused. "One of your buddies?"

"Yeah. One of my locksmith friends who specializes in car openings."

Joe was incredulous. "Wait a sec," he said. "Do you mean to tell me that you can get into a bank vault, but you can't get into your van? You're kidding, right?"

I smiled. "I'm serious."

Joe stuck around and watched an expert bail out a dabbler.

NEVER SAY NEVER

I have drilled many bank vaults in Chicago, but the most memorable is one I *didn't* drill. Built in the 1930s, the Royal Savings Bank on the city's south side still sparkled with marble, brass, and stainless steel. It was a reminder of a bygone era — they don't make 'em like they used to.

In the basement were two gigantic vaults. They had the combination to one and used it every day. The other one had been locked for a decade. Auditors demanded that the unused vault be opened and its contents inventoried. But no one could remember the combination.

I was staring at a gleaming old Mosler, seven feet tall, nearly four feet wide, and more than a foot thick. By any measure, a tough opponent. The vault door was held locked by twenty-four massive round bolts, and all sorts of surprises were hidden in its thick layers of Brooklyn chrome steel. It was David versus Goliath, or at least David versus Mosler, which is pretty much the same thing. (Notorious bank robber Willie Sutton once gave up after spending hours attacking a safe, allegedly leaving a note that said, "Mosler — too tough.")

My magnetic drill press was on the door, motor running, drill bit about to contact steel when the branch manager came clickety-clacking down the long flight of stairs. "Wait!" she shouted, trying to be heard over the whir of the heavy-duty drill motor. She was waving a little piece of yellow paper in the air.

I was too distracted to notice, but Ken Nowikowski, the onsite Diebold technician, saw her rushing toward us. He gently nudged my shoulder as she approached the vault. "Ethel might know the combination!" she said, out of breath, and handed me a post-it note with numbers scribbled on it.

Ethel had retired a decade earlier. She was on vacation when they called to ask about the combination. She returned that very day, listened to her messages, and immediately called the bank. I turned off my drill, dialed the numbers, and the vault opened.

For me, this experience underscores a valuable insight: just because we don't have an answer today, doesn't mean we won't tomorrow. Remain open. Reject dogmatic certitude about nearly everything.

And never say never.

Chapter Thirty

Thinking about Tomorrow

I'M SITTING AT MY OLD OAK DESK. EVERYONE IS IN BED. THE SILENCE AND solitude are conducive to planning, and for the past hour I've been going over Plan A and Plan B. Both are solid. My bags are packed and in the van; I'm ready for an early morning departure. Everything looks good, but still, I can't quite shake the downer.

Tomorrow's job is the result of a tragedy.

On the lower level of the most famous residence in Minnesota sits a massive Mosler bank vault. For thirty years, its owner controlled access by keeping the combination to himself. There are rumors, some of them blatant fabrications spun up by an overzealous media,[1] but no one seems to know for sure what's inside. What we do know is that the vault's owner, Prince Rogers Nelson — known to the world simply as "Prince" — personally closed and locked the three-ton door awhile back, sealing its contents from the world.

My only task for tomorrow is to open it.

Sunday

Chapter Thirty-One

Paisley Park

THE CHECKERBOARD BELOW BEGINS TO FADE AS WE GRADUALLY ASCEND into a dense raincloud. I watch the flashing white light on the wingtip slip from view, followed by most of the wing a few seconds later. Fully immersed now, high in the heavens above Portland, visibility is measured in mere inches.

Newcomers to the Pacific Northwest often lament the rain, but I suspect our gray skies take the greater toll on transplanted psyches. Maybe it's because I was born here, but overcast is my favorite sky. There's less need for greasy sunscreen, more time for reading, and things just smell fresher, crisper, *better* when they aren't burnt. A little vitamin D goes a long way.

We poke through the cloud layer and are greeted with early morning rays. I lower the shade and hear the double chime. Not in the mood for a movie or book, I pull out my phone and cue up a mellow playlist, close my eyes, and lean back against the headrest.

This is not going to be an ordinary day.

Prince graduated high school the year after I did and was still a teenager when he signed his first recording contract. Our generation cheered the rise of this gifted songwriter who sang every part and played every instrument on his debut album.

An electrifying performer, Prince quickly vaulted from mere stardom to superstardom, where he stayed for nearly four decades. But now, impossibly, he's gone — an uncomfortable reminder of the thin and fragile line each of us is destined to cross.

Some artists build recording studios in their homes, but, as I learned, Prince did the opposite. He assembled multiple state-of-the-art studios under one roof, along with a full-size sound stage and concert hall. Prince liked the place so much that he moved in and made it his home. He called it Paisley Park.

We pull up behind a white sedan waiting at the gate just off Audubon Road. A man wearing a black uniform and an official-looking badge is speaking with the driver and checking ID. On either side of the gate is a wall of purple. As we inch closer, I see that it's actually a tall, chain-link fence, hardly visible through all the tributes left by grieving fans.

I watch a young couple tie up a large purple ribbon with a note. I can't make out what it says, but the flowing calligraphy is clearly purple, as are many of the thousands and thousands of items hanging from the fence or resting on the sidewalk around the perimeter of the property.

"Traci, this is amazing."

She nods slowly. "It really is. He was loved by so many people."

Once Prince's personal attorney, Traci now works for the firm handling the estate. She called to set up the vault opening, took care of all the paperwork, and picked me up at the Minneapolis airport about thirty minutes ago. She even bought the Starbucks Americano that I downed as soon as it was cool enough to drink.

The gate guard waves us through, and Traci parks her sporty Volkswagen near the entrance. Stepping out of the vehicle, I get my first clear view of the enormous white aluminum and steel building where so many iconic songs were recorded.

"What can I carry?" she asks, surveying the cases. When she talks, she smiles. That's a quality I wish I possessed. And what a gem — coffee *and* a helping hand. I thank her and pass over my aluminum scope case.

I slide my computer case down around the Beemer case's telescoping handle, grab the other case, which houses my portable drill press, and follow Traci inside. We walk past another guard and through a reception area into a large atrium, where I can see a small slice of the fabulous fifty-five thousand square feet Prince lived and worked in for so long.

The gate guard at Paisley Park checks each vehicle before granting entry. We are second in line.
Source: Author photo.

The carpet is a light steel blue, dotted with large yellow stars and crescent moons. My eyes are drawn to the multicolored decorative columns all around the room and the light purple walls in the background, lined with entertainment industry awards.

Along the far wall are a number of doors, each adorned with a different drawing or painting of Prince, either in a pensive pose or holding a guitar. Two large purple sofas face each other, bathed in natural sunlight,

The fence next to the front gate at Paisley Park is hardly visible through all the fan mementos.
Credit: Nancy Rosenbaum

with a rendering of Prince's iconic symbol inlaid on the tile floor between them. Photographs don't do justice to Paisley Park or its vibe. This is a hip and happy place on a most unhappy occasion.

I'm jolted out of my tourist trance when Traci begins introducing me to the many people who now surround us — lawyers, bankers, and a world-renowned archivist from New York named James.

Music insiders have often remarked on Prince's almost superhuman work ethic, and it's entirely possible that many unreleased studio recordings are inside the vault. Warner Brothers stoked this rumor long ago with the following verbiage on the promo jacket for Prince's 1999 album, presciently titled *The Vault . . . Old Friends 4 Sale*: "It has long been known that only a small portion of Prince's remarkably prolific creative output has been released."

If that's true, and if I succeed in getting the vault door open, James will take possession of those precious tapes and preserve them for later mastering and potential release. Fans across the world would love to hear Prince's voice and guitar again, especially on new material. Time will tell.

"Shall we?" he asks, turning to lead the way toward a narrow set of stairs where we funnel into a single line and head down to the lower level. At the bottom of the stairs, I set the Beemer case on its wheels, raise the handle, and follow James toward the light emanating from a room at the end of the hallway.

Three DEA agents meet us at the doorway to the vault room. I know from news reports that they've been here since the Carver County Sheriff discovered prescription drugs onsite and asked for assistance. Michael Jackson's doctor went to jail; I'm guessing Prince's doctor is sweating bullets right about now. One agent steps forward, introduces himself, and instructs us on the obvious protocol: no one sets foot inside the vault until they complete their search.

He doesn't know it, but this is SOP in my line of work. Whenever I open a safe or vault with unknown contents, I step aside and give the responsible party first access. The reveal is theirs, not mine.

I nod. "Of course."

We walk through the doorway, and there along the left wall is my opponent — a gleaming, Mosler American-Century bank vault. The door measures six-and-a-half-feet tall by three-and-a-half-feet wide. More than eight inches thick, it weighs in at nearly six thousand pounds.

In the center of the stainless steel door are two black and silver dials, stacked one above the other about a foot apart. To their left is the vault's only welcoming feature: a black and silver captain's wheel, always inviting the curious to give it a spin. By the looks of the fresh fingerprints all around the edge of the handle, someone has tried multiple times.

To the right of the dials are two large hinges, each one as thick as a fist and as long as an arm, neatly concealed under a black, full-length cover that is slightly askew. Someone has pried on it with a large screwdriver or pry bar, perhaps the same person whose fingerprints are on the handle.

With its high-contrast two-tone motif, Mosler's American-Century is as aesthetically pleasing as it is imposing; it's an attractive monster.

This Mosler American-Century vault door has evidence tape bridging the door and jamb.
Credit: Carver County Sheriff's Office.

I take a step back, and my eyes are drawn to the wide strips of tape crossing the door and jamb. Evidence tape. It bridges the door and jamb in multiple places, and the door cannot be swung open without breaking those seals.

Like any serious safecracker — nearly all of us are blissfully afflicted with a touch of OCD — I have known the make and model of Prince's vault ever since one of his friends posted a photograph on the web many years ago. The purple collage above the dial in the original photo is now gone, but this is unmistakably the same door.

Mosler was America's premier safe and vault maker for more than a century before their tenure came to an abrupt end in 2001, and the American-Century was the last of their truly innovative doors. Melding old-world craftsmanship with new-world security, it was designed to withstand drills and torches, to repel sophisticated attempts at entry and brutish forms of attack as well. To my knowledge, no burglar has ever successfully breached an American-Century.

I remove the electronic stethoscope and headphones from the front pocket of my main tool case. Like a doctor checking a patient's heart and lungs, I move the microphone from place to place on the vault while rotating each of the two dials, listening intently to every sound emanating from deep inside the door. The chill touch of the stainless steel feels good in this warm room. Both dials spin freely and are rotating their respective tumblers smoothly, with no indication of an underlying problem. This is good.

I'm betting Prince was far too meticulous to use any of the standard factory or default combinations, but I try them anyway. To no avail.

Trade secret: for ease of dialing, many American bank vaults have combinations set mostly on 0s and 5s. But before you rush to your local financial institution at midnight and start dialing combos, you should know five critical facts.

First, the vast majority of bank vaults use combination locks with four tumblers rather than three, which means there are one hundred million possible combos. Even if three of the four tumblers are set on 0s and 5s, the number of potential combinations is still in the millions (because you don't know which tumbler *isn't* set on a 0 or 5). Second, nearly all bank vaults use *two* locks rather than one. While most banks set both locks on the same combo, some do not, and there is no way to tell. Third, even if you do stumble onto the correct combo, and even if the vault is set to single custody, *the timelock will prevent the vault from opening in the middle of the night.* Fourth, even if you were to miraculously dial open the combo locks, bypass the timelock, and open the door, the alarm would go off and you'd be caught just as you discover, fifth and finally — the bank's cash is stored in yet another locked safe inside the vault.

Don't do it!

Chapter Thirty-Two

The Mousetrap Relocker

I HAVE MADE A CAREER OUT OF FINDING THE WEAK SPOT IN JUST ABOUT every make and model of vault door. There always *is* one, and the manufacturers themselves are usually unaware unless it is explicitly brought to their attention. Not so in this case. The vulnerable point in an American-Century is tiny, about the size of a frozen pea. But Mosler's engineers figured it out and placed a fusible link in this area and ran a cable up the door to a fiendish relocker.

If I get the fusible link too hot or drill it in half, its cable will release and everyone in the room will hear the loud SNAP of a diabolical, spring-loaded relocker firing. This relocker is designed like a large, powerful mousetrap, and when it fires, it prevents the vault's handle from turning, even *after* the combination is dialed.

Every single American-Century vault door that I have drilled at banks and credit unions around the country has caused me to grow a few more gray hairs, worrying about heat and friction. I've never fired a mousetrap relocker. But I need to be on my A game and not let this turn into a cage fight, or I'll lose spectacularly.

A gentle hand is on my shoulder. "Dave?" I turn to face one of the gentlemen from the bank handling the estate. He gestures to a video camera on a tripod. "Do you mind if we tape the process? We're trying to document everything as thoroughly as we can."

There's no polite way to refuse. "Sure, no problem," I reply, as images of fusible links and mousetrap relockers fly across the high-definition screen in my mind's eye. I force out the negative thoughts with a vision of this magnificent vault door swinging open.

I notice that the items hanging from the wall on either side of the vault are covered in clear plastic. Curious, I glance around this rectangular room and see that every plaque and photograph on every wall is covered in plastic, as are the boxes on the floor. That's a *lot* of Saran Wrap! I look over at James. My face must convey puzzlement because he anticipates my unspoken question.

"We weren't sure what to expect when you got here, Dave, so we covered everything just in case there are explosions."

I smile inside but remain impassive. People often think this and are surprised to learn that a safe or vault will be completely usable once it's properly opened and repaired. "No explosions, James. We just need to drill a small hole through the door in exactly the right spot."

"Okay," he replies, "that's good. Minimal vibration then."

"That's right."

I begin my predrilling ritual. I usually lay my cases to the left and right of a vault door for easy access. But given the number of bodies packed into this room, I want a barrier that people will be reluctant to cross. So I ask the crowd to back up a little, to protect both them and my tools, especially the delicate medical scopes. It's a friendly group, and they comply.

About six feet back from the vault, I arrange my cases in a half circle. I then pull out two tool rollups packed with drill bits and unroll them to fill in some of the gaps between cases. It's not an impermeable barrier, but it works.

I remove a few necessities from my main tool case: a short stack of large shop rags, a measuring tape, and a roll of painter's tape. I tear off a short strip and, from memory, stick it on the vault door, then carefully measure out my preferred drill point and mark the sky blue tape with a single black Sharpie dot.

I get down on my knees and start taping shop rags to the floor in front of the vault, and realize that someone has already laid a blue plastic tarp over the carpet. That's great, but I tape down a couple of rags anyway, directly below the Sharpie dot. They're easy to roll up and toss when the job is done.

Catching right on, James asks, "Are those for catching drill swarf?"

I'm impressed. Swarf — the spiraling metal remnants created by the working end of a drill bit — is a word you don't often hear. "That's right, James. I don't like leaving a mess behind, and this simplifies cleanup."

"I have a Shop-Vac here, Dave. Would you like me to use it while you drill?"

I'm not sure how well it'll work, but I like the idea. "That would be great. Thank you." He walks to the corner, rolls his Shop-Vac in my direction, lifts it over the tool cases, and parks it next to the vault.

I double-check the location of the Sharpie dot, then pull out my phone and open WWF. Valerie won our most recent game; I almost pulled off a crazy comeback this morning but fell a few points short. She launched a new one while I was in the air and played VOIDING — a seven-letter word. *Nice start, hon!* I have six frikkin' vowels and an L, so I play VIOLA off her V. We've each gone once, and I'm already behind by more than forty points. She's about to pull even with me in all-time wins.

I triple-check the drill point, stand and turn, and catch myself staring straight into the video camera. *Look away, Dave!* Jeez, I wonder how long it takes actors to overcome the natural urge to eyeball the lens. Note to self: the tripod is about ten feet behind me and four or five to my right. Avoid the appearance of amateur hour by not glancing in that direction again.

Two men in matching work shirts emerge from the crowd. Ah, it's Dick and Mike from Midwest Safe Company. I wave them over. These guys have been in the business for a long time. We shake hands and pull up two chairs for them to complete the half-moon barrier.

On long-distance jobs, when I'm not working with bank service techs like Hitch and Robert from Diebold, I like to bring in a local company to help with the repair and set a new combination. Not out of unmitigated altruism, but to avoid putting my warranty on safes and vaults that are thousands of miles away. Once I get the vault open, Dick and Mike will set a new combination and make sure the bank people are familiar with the dialing process.

I reach into tool case number two and remove the components of my portable drill press: base, compressor, and drill stand. Anyone who has ever drilled by hand knows how difficult it is to hold a motor absolutely straight. This is doubly true when the target is small and far away. A drill press holds the motor straighter than human hands can and provides maximum precision. I don't always take mine on airplanes — it's a little heavy and I'm a little lazy — but it made the trip with me today to ensure a straight hole through a thick door.

I plug in the compressor and run its air hose over to the base, which I lift and press lightly against the vault. This base is a vacuum pad with a footprint of about one square foot, made from lightweight aircraft aluminum, and with a soft gasket around its perimeter. I start the compressor, and the airtight gasket gradually flattens as the base sucks down against the stainless steel.

Next, I position the drill stand and its onboard motor over the threaded mounting posts on the base. Finally, I snug up two knurled black knobs, each the size of a small donut. The base and drill stand are now an integrated unit. I loosen the keyless chuck on the Fein 638 drill motor and insert a 5/16" carbide bit, making sure it's aligned with the Sharpie dot. Setup is complete — we're ready to rumble.

Breathe in, breathe out. *Okay, let's do this.*

CHAPTER THIRTY-THREE

Drilling the Vault

I DEPRESS THE TRIGGER ON THE FEIN, AND THE CARBIDE BIT BEGINS TO spin. In two seconds, it's idling at eleven hundred RPMs, my preferred speed through the first few inches of material.

I look over at James. "Ready?"

He nods, flicks the switch on his Shop-Vac, leans over, and places the end of the hose against the blue tape about an inch below the Sharpie dot. The whine of his vacuum almost drowns out the whir of my drill motor. This is going to be a raucous concert.

The drill stand has three spoke handles to move the motor in and out. I pull gently on one, pushing the bit against the door, through the dot, and into the stainless steel. Chips immediately begin to fall — not onto the shop rags taped to the floor, but into the hose James is holding. This is amazing, and a first for me. The mouth of the hose is like a black hole, and the chips are like light beams being sucked in, unable to escape. I look over and grin, nodding my head to an inner beat. *Why didn't I think of this?*

The Fein's cooling vent is blasting air right in James's face. His silver hair is blowing back like he's riding a motorcycle without a helmet. But he doesn't waver, and the hose stays put. He's a trouper.

We transition from stainless to mild steel and then to high-PSI concrete in this composite door. Chips and dust disappear into the black hole until the sound of the motor rises half an octave and its speed increases. Progress has stalled. We are at the hardplate.

I remove the dull bit and chuck up another, whose carbide tip has been custom-sharpened to a knife's edge. I rotate the speed selection knob on my motor to High and pull on a handle spoke. The bit spins against the hardplate and goes nowhere. I pull harder. Still nothing. Maybe the angle on the tip is wrong. I remove the bit and replace it with another, and pull hard and long enough on the spoke to smell the Fein's windings getting a little too hot. But the result is the same — we're getting nowhere. *Why?*

It's time to inspect the hardplate, so I reach into my aluminum case and remove a footlong, straight-view scope. I peer down the hole, expecting a smooth plate, but this stuff is rough like cookie dough — carbide chips in a bronze matrix. *Dammit!* I recognize the hardplate. Mosler was so proud of this diabolical plate that they reversed the company name to give it a memorable moniker. M-O-S-L-E-R became R-E-L-S-O-M, and Relsom is redrum to penetrate.

This isn't what I'd hoped for, especially today. I've drilled a lot of American-Century vault doors, and none of them contained Relsom. They either had smooth hardplate or a hybrid that didn't cause me too much trouble. I wonder if Prince talked Mosler into a special hardplate-from-hell upgrade. We'll probably never know.

There are old- and new-school methods of penetrating Relsom. The old-school method is a punch and drill attack: we insert a pointed, hardened punch and pound on it with a mini-sledge until we move the carbide pieces away from the center of the hole. Then we switch to high-pressure drilling. Then back to punching. Then back to drilling. Punch and drill and punch and drill until we have a hole all the way through the Relsom. There's just one problem: high-pressure drilling generates tremendous heat, a fact of which Mosler was all too aware — thus, the fusible link attached to a mousetrap relocker.

The new-school method is less brutish. I reach into the Beemer case for a bank bag containing an assortment of small metal tubes that look like drinking straws with diamond particles silver-soldered all around the tip. They'll penetrate almost *any* hardened material and provide safecrackers with a near superpower. Diamond bits should come with a cape.

The hole to the Relsom is 5/16″ in diameter. We're stepping down to a 1/4″ diamond bit, and after we get through the hardplate, we'll step down one final time to finish the hole into the lock.

The key to success with diamonds is also the key to avoiding a problem with the fusible link — keep heat and friction to a minimum, otherwise the diamonds will be torn off the tip of the bit. We minimize heat and friction by drilling at a relatively low speed with light-to-medium pressure, and by keeping the diamonds well lubricated. We're not going to anneal this material; we're going to slowly *grind* our way through it.

I chuck up a 1/4″ diamond bit, turn the knob to put the Fein back in low speed, ease the bit into the hole down to the Relsom, and start drilling. A slow slog lies ahead.

After a few minutes of watching paint dry, I notice James looking at me. This time the puzzlement is on *his* face, and I explain: "Diamond drilling produces mostly sand and dust, and it all stays in the hole. There won't be any more chips or drill shavings until we get through the hardplate. You can take a break." He nods, pulls the hose away, and turns off the machine.

I remove the bit to examine the tip. It's in great shape — the diamond particles are intact. Time to relube. From my vest pocket, I withdraw a small container resembling lip balm, open it, and smush the tip of the bit into the waxy pink lubricant. This stuff comes from the aerospace industry and works like magic to reduce heat and friction.

Grinding begins anew, and I'm pleasantly surprised to see the little hole saw creeping deeper and deeper into the Relsom. With a bit of luck, we'll make it all the way through with only a single diamond bit casualty. Progress continues unimpeded until the bit jumps a little and the motor revs in idle. Contact is made deep in the door, and a horrible metallic screeching fills the room.

What the heck? I pull the bit out of the hole and examine the tip. The diamonds are pretty much gone, but the bit's previously hollow center is now filled with a long piece of Relsom. *No way!* I peek in the drilled hole with a straight-view scope and can hardly believe my fantastic fortune: there's a quarter-inch hole all the way through the hardplate! We're into a slight air gap, with nothing but mild steel on the other side.

I toss the diamond bit down onto the shop rags. It performed admirably, but now it's toast. From the larger tool roll, I grab a footlong 3/16″ high-speed bit and slide it down the hole against the mild steel with the

Fein turning at low speed. It begins cutting immediately. James turns the Shop-Vac on just in time for several long steel spirals to drop into the black hole, and the bit jumps forward into dead air. The drilling is done.

I release the trigger on the motor, and James flicks the switch on the vacuum. The whir and the whine fade quickly from a hundred-plus decibels to zero, and I'm struck by how palpably quiet it now is in this crowded room. I pause to take stock. There was no loud SNAP, which means the fusible link is intact and the mousetrap relocker hasn't fired. *Right?*

There's only one way to know for sure. I need now to look sideways rather than straight ahead, so I grab a right-angle scope from the case and slowly insert it into the hole. I realize that I'm holding my breath. As the scope passes into the air gap, I point it straight down and see that we just barely squeezed by the fusible link. We cut it close, but you *have* to on these doors.

Still holding my breath, I rotate the scope about twenty degrees clockwise. The view couldn't be better — all four tumblers and the lever are clearly visible. Target acquired, I exhale loudly. It's all downhill from here.

In the remake of *The Getaway*, Alec Baldwin drills deep into an armored vault, slides a scope into the combination lock, and transmits the image to a monitor. The viewing audience sees what he sees. Except for the trivial fact that it's the wrong model of combination lock for that make of vault door, the scene is both uncannily accurate and fun to watch.

The monochromatic view Baldwin has on his monitor, though, pales in comparison to the delectable eye candy I'm seeing through my scope. As I turn the spindle, the white thermoplastic tumblers look like little grindstones, rotating this way and that until they are properly aligned under their mating piece, itself smaller than a scoring peg on a cribbage board. Each time I pass zero, the bright brass lever wiggles back and forth, demanding — and getting — my attention.

On the last turn of the spindle, I observe the stainless steel nose hesitating momentarily and then dropping into its notch. I linger for a few seconds to watch the chunky brass lock-bolt slowly pulling back. There is a faint "tick" as it reaches full retraction inside the lock.

This vault is all but open. I feel my opiate receptors being flooded by chemicals whose natural production in the body must be counted among life's happiest miracles. As casually as I can manage, I step up, grab the captain's wheel, and give it a clockwise tug. It doesn't budge.

In a blink, my euphoria is gone.

All right, Dave, slow down and think it through. The fusible link is intact. You know this for a fact because *you saw it with your own eyes.* This means the mousetrap relocker didn't fire. So why won't the handle turn?

There *has* to be a logical explanation. And it hits me: most bank vaults that are hinged on the right side have a handle that rotates clockwise to open, but the American-Century is an exception to the rule. This handle rotates counterclockwise.

I gently rotate the captain's wheel counterclockwise until it stops. Victory! I tug outward, but the door stays solidly locked in the jamb. It won't move. I tug again with the same result. The door won't swing open. *This is impossible.*

I close my eyes for a second and take a breath. Mr. Corey's aphorism comes immediately to mind. I'm trying to not let them see me sweat, but two dozen pairs of eyes are boring a hole in my back. If I don't get this figured out fast, I'm going to royally embarrass myself by turning beet red in front of God, the video camera, and everybody in this room. If ever I needed a vault door to swing smoothly, it's this one.

I pull on the handle again and hear a tearing sound that doesn't belong. What *is* that? My eyes search for the source over near the edge of the door. The evidence tape! It's holding the door shut, and a tiny bit is peeling away with each tug. I pull *hard* on the handle this time, maintaining consistent outward pressure, and the strips of tape slowly peel away from the jamb and wrap around the edge of the door as it slowly creeps partway open.

When there is enough space between door and jamb to ensure that we are home free, I stop pulling and turn to the agents. "Give me a minute, guys, and it's all yours." I squat down and roll up the rags that would usually be covered with drill swarf. There's almost none, which makes me smile inside.

I stand up and pull gently outward on the handle one last time. With no more resistance from the tape, the exquisitely balanced three-ton door swings all the way open. And in they go, one after the other in their matching windbreakers, disappearing into darkness until somebody finds the light switch.

My shoulders relax, my heart starts to beat regularly again, and, all of a sudden, it doesn't feel quite so hot in here. It occurs to me that the last person to open this vault was Prince himself. That's crazy.

CHAPTER THIRTY-FOUR

Inside the Vault

I GLANCE OVER AT JAMES, WHOSE EYES ARE LASER-FOCUSED ON THE vault's interior. I'm programmed not to look, so it takes me a second to force my gaze beyond the doorway. Wow, it's bigger than I expected. Approximately twenty feet wide by forty feet long, this vault room is on the large side, even by bank standards.

Despite its size, this is the least roomy interior I've ever seen, because it's lined wall to wall with industrial steel shelving units. There are sixty or more of them, and most are jam-packed with reel-to-reel tapes on all five shelves. My goodness, there are thousands and thousands of tapes here! James will be putting in some serious overtime to archive all this material.

Dick and Mike walk over. "That went well," one of them remarks casually.

"Yeah, we got a little lucky," I reply. "It could have gone south at any time."

We remove the stainless steel panel from the back of the vault door and lean it up against the wall. The two combination locks are visible, as are the timelock and many components of this unique locking mechanism.

"Where did the hole come out?"

I'm not sure who asked. I stand off to the side and point to the tiny hole, a whisker above the fusible link and its stranded, stainless steel cable. That cable runs under the locks, then around the handle, and finally up and over to the mousetrap relocker near the edge of the door.

"Whoa, Dave. Good thing you weren't using a great big, red-hot drill bit."

Indeed.

Inside the vault are row after row of library shelves jam-packed with recording tapes in their labeled boxes.
Credit: Carver County Sheriff's Office.

I begin patching the drilled hole while Dick and Mike take the combination locks apart to set them on new numbers. They'll try the combinations a little bit high and low to ensure that the numbers are properly centered and make sure the representatives from the bank can dial open the vault.

Because they are experienced professionals, Dick and Mike will do this with the door locked *open*, in compliance with the first commandment of safe and vault work: ***Never test a new combination with the door shut and locked.*** That way, if the combo doesn't work, the problem can be easily fixed, and the door won't have to be drilled again.

Most of us learn this lesson the hard way. Once. Then we never break that commandment again. Experience is a brutal teacher. The test often comes first, *then* the lesson.

One of the big, burly agents walks out of the vault and heads my way. "We're done, except for one thing," he says. "There's a little safe just inside the vault. It has a key slot, not a dial. Can you open it?"

All my special tools for keylocks are in my safe at home. "Sure," I reply, hoping it's an easy one. "Let's have a look."

I follow Burly through the doorway. We turn left, and I spot the ventilator on the jamb next to the door. It looks like new. All modern vaults in the United States are required to have emergency access to oxygen, in case a person is trapped inside (which happens every so often).

Vault makers cleverly disguise their ventilators in a variety of ways. The American-Century uses a large tube through the foot-thick concrete wall, connected to a fan on the inside. The tube isn't visible from the outside because it's hidden under a dial that isn't connected to any tumblers. This fake dial is just for looks — it covers the inch-and-a-half-diameter hole through which the fan in the ventilator sucks air. You can't see the tube on the inside either, as it, too, is covered by another fake dial.

Ventilators have been required in bank vaults for more than half a century, and in that time we have lost very few people from being trapped inside. Even a soft libertarian like me can acknowledge that *some* regulations are eminently sensible.

Burly and I approach a table on which rests a small Sentry keyoperated safe. This is a huge relief, as these are among the easiest of easy. It yields the instant I insert a tool, making me look like an expert locksmith, which I'm not.

"You can open it now. It's unlocked."

I reflexively look away and notice a computer sitting on the table next to the safe. *Why would Prince put a computer **inside** the vault?* To the right of the computer is a piece of copy paper with some text across the top that catches my attention. It reads, "Mr. Vault Guy." Hey, that's *my* nickname!

I glance down the page and realize these are instructions. The *computer* is Mr. Vault Guy, and his purpose is to aid the user in finding particular tapes among the many thousands on the shelves here.

I turn to exit the vault and cannot help but glance at the cardboard boxes the tapes are in. Sure enough, there are numbers or letters (or in some cases both) written in black felt pen on every single box that I can see. I get it now. Type in a search term and Mr. Vault Guy tells you which box to hunt down. It's Prince's version of the Dewey Decimal System. Genius!

Mr. Vault Guy!
Credit: Carver County Sheriff's Office.

I walk back over to the vault door and resume work, smiling inside. I'm putting the final, cosmetic touches on the repair when the agents file out of the vault. Their search is complete. They found nothing, as expected, but they had to be sure. The agent in charge gives the go-ahead to the people from the bank who nod at James. I shake his hand at the doorway and thank him for his help. I don't envy the job that lies ahead of him, but I do have a new and novel use for my own Shop-Vac back home.

My task complete, I shake hands with Dick and Mike, thank them for working on a Sunday evening, and leave my cases in the hallway against the wall. I turn to canvass the room and spot Traci along the far wall, chatting with another lawyer. She sees me heading her way and smiles, Effortlessly. "Are you done already?"

"I am."

"And your guys?"

"Almost. They just need to get with the bank people and go over the dialing process. But I'm not involved in that."

"So, are you ready to go to the airport?"

"I am."

We head for the hallway. Unprompted, she reaches down and grabs my aluminum scope case and leads the way to the stairs. When we get to the top, I lower my tool case back onto its wheels, slide my laptop case over the handle, and follow her around objects and people. We pass the security checkpoint out the door to a cool breeze and her car.

Traci waves to the gate guard, turns right on Audubon, then right again on Arboretum Boulevard. We have a sweeping view of Paisley Park, including the famous egg-shaped building behind the main complex. The fence along Arboretum must be five or six hundred feet long, and it too is adorned with notes and flowers and purple fan mementos all the way to the end of the property. I can't help but stare and snap a photo through the car window.

"Surreal, isn't it?" she asks quietly.

"Yeah," I reply, nodding slowly. "This whole thing is like a bad dream. But what a beautiful gesture from so many fans. I've never seen or heard of anything quite like this."

"Neither have I."

I notice a familiar round sign on the block ahead. Traci sees it too. "Hey," she asks, pointing to the ubiquitous mermaid, "How about a Starbucks for the road?"

I usually avoid coffee this close to boarding a plane, but it sounds great and will keep me from dozing off on the flight home.

"Yes, please."

The fence along the highway isn't as densely packed with fan memorabilia as the fence along Audubon Road, but it's still a once-in-a-lifetime sight to behold. In the background are the Paisley Park buildings.
Source: Author photo.

At the airport, Traci steers her car to the curb as I down the last of my Americano and reach for the door handle. "Hey, before I forget," she says, "I want to thank you for coming to Paisley Park and doing such a professional job. You just made a huge headache disappear for a whole bunch of people."

"It's nice of you to say that, and thank *you* for being so great to work with, and for the coffee. Yeah, the vault opening went well, but Murphy could have shown up at any time and caused us all sorts of problems."

"You'd have missed your flight!"

"Well, it's happened before, and it'll happen again. Sure glad it didn't this time, though."

"For sure."

We wave goodbye, and Traci pulls into traffic while I head inside the terminal.

What a lady. What a day.

Chapter Thirty-Five

Coda

I HEAR THE FAMILIAR DOUBLE CHIME, OPEN MY EYES, AND NOTICE THAT I'm seated between two kind souls who have yielded the inside armrests. I uncross my arms and extend my elbows just far enough to be comfortable. There isn't a graceful way to thank them, so I wear a smile for a while.

I have opened countless safes for the estates of people who have passed, and I usually have no problem detaching myself from the situation, doing my job, and moving on to the next one. But I think Paisley Park is going to stay with me awhile.

The young lady in the window seat pulls a dark blue three-ring binder out of her backpack. She's probably in high school or college. On the cover of her binder are two fish stickers I've never seen together. The upper one contains the Christian symbol IXOYE, the lower one DARWIN. I'm tempted to ask if she believes in God-directed evolution rather than random mutation and natural selection, but she's preoccupied with her studies, and I don't want to be that annoying seatmate who insists on a conversation over obvious hints to the contrary.

The debate over the origin and proliferation of life has raged for a century and a half and shows no signs of abating. I struggle with this issue and the cluster of related ones on both the micro and macro levels. They seem to pit intuition against intuition in an endless tug of war.

Sometimes when I recline in my amateur philosopher's armchair and ponder our finely tuned universe, or marvel at the complex nanofactory inside a single cell, it seems wildly implausible that it all could have happened purely by chance. During these moments, it seems likely that behind the appearance of design stands an actual Designer, or perhaps an entire team of designers. If so, we are the beneficiaries of a process we can only dimly comprehend.

But other times, when I witness or feel the raw emotion of personal tragedy, or confront the problem of suffering generally, I have serious doubts. During these moments, it seems likely that behind the seemingly miraculous appearance of life in the cosmos stands nothing but staggeringly improbable odds. If so, we are supremely lucky ticket holders in the mother of all lotteries.

The possibilities here aren't infinite, but binary, and they perfectly mirror my vacillating intuitions: either we're here by design or we're not. But which is it? I have close friends on both sides of this issue, but I'm stuck in the muddled middle — *I don't know.* Just once, though, I'd love to break the stalemate and feel the certitude of either side, if only for a moment. It must be like a drug.

I love challenges, like figuring out how to crack tough safes and attempting to solve life's most perplexing puzzles. Why *is* there something rather than nothing? What existed before the Big Bang? If there is a Designer, does He delight in being the greatest Hider in the history of hide and seek? And, of course, the questions of great philosophical import: which of my techniques would be most effective on the gigantic vault at Fort Knox? Are there bank vaults elsewhere in the universe?

I know the answer to only one of those five questions. The rest are mysteries, and that's okay. Not knowing something is half the fun — it provides motivation for the thinking and learning that give rise to much of what we value in this life.

Floating north in this aluminum eggshell, thirty thousand feet above terra firma, I may not know much about ultimate destinations. But I do know where I'm headed tonight: I'm going home, where I belong, to be with my family. To crack more safes, read more books, and ponder life's mysteries for whatever time remains.

Deplaning seems slower than normal. I watch out the window as bags are lifted off the endless conveyor belt and placed onto the cart for transport to baggage claim. They are different sizes and shapes, most of them ballistic nylon, some hard plastic, and a few aluminum. Some, like mine, have a brightly colored ribbon around the handle; most do not.

Like safes and vaults, each bag is safeguarding somebody's stuff. If there were no personal property, or if human nature had no dark side, I'd be out of a job. Given the size of those ifs, though, I suspect the world is going to need safes and safecrackers for a while yet. Maybe longer.

Heading home in the van, I merge onto the freeway and shuffle an old Prince playlist I haven't heard in years. What unbelievable range he had, both vocally and lyrically. Because of his talents as a singer, songwriter, arranger, and entertainer, it's easy to overlook Prince's otherworldly musicianship.

I'm a lifelong guitarist who spent a fair amount of time onstage and in the studio. But Prince was on another level entirely, one shared with only a handful of groundbreaking players across musical genres — Jimi Hendrix, Jimmy Page, B. B. King, Lenny Breau, Chet Atkins, James Burton, and a few others.

I pull into the driveway but can't bring myself to turn off the ignition during his solo in the middle of Purple Rain. It's fantastic. In my mind, I see those rows and rows of studio tapes in the vault, and I can't help but wonder what guitar riffs we may yet hear. Prince may have departed this world, but his artistry and body of work will live forever.

If I open the front door, the dogs will bark and wake Valerie and the kids. At this time of night, it's best to go through the garage. After the guitar solo, I hit the button to open the overhead door and briefly debate whether to put my tools away. I'm tired, but, with no jobs scheduled for tomorrow, I really should. In the morning, I'll sharpen the drill bits that were dulled this week, and prepare for the next.

I walk around to the back of the van and lift out the suitcase that houses my tools. Man, fifty pounds is a lot heavier now than it was this morning. I wheel the cases inside the overhead garage door and dial the combination to my largest (but not heaviest) safe. Then I slide my aluminum scope case onto its dedicated shelf and reach down to lift the suitcase by its handle. My eyes focus on that tattered red ribbon, and I pause.

Along the wall to my right is a large shelving unit containing sundry household items. Every home has its organizer, and in ours it's my wife. Valerie loves sorting stuff into plastic bins and handwriting category names on peel-off stickers. The only one I access with any frequency is "Light Bulbs." The others are largely a mystery, except "Wrapper," which I open for birthdays and Christmas.

I slide the lightweight bin out from the shelf and remove its lid. Inside are a few rolls of wrapping paper, and a bunch of curling ribbon rolls in a variety of colors. I see red, green, yellow, white, pink, two different shades of blue, and purple. *Purple.* I never noticed that one before.

It's perfect.

POSTSCRIPT

OKAY, LET'S TIE TOGETHER A FEW LOOSE ENDS. I WENT TO ALBUQUERQUE the following week and opened that malfunctioning vault for Bank of America. The job went perfectly, but BofA is yet to change their policy, which still calls for a gigantic hole to be cored through the wall. Every vault I open for them is hit and miss as it climbs the chain of command. At this point, I'm convinced that there won't be a new policy until the old guard retires.

Friends and acquaintances at Diebold have been retiring all around me. Perrill Smith from Portland and Don Spenard from Seattle recently bade farewell to the big D, as did Mike Hitchcock from Las Vegas, Robert Coleman from Salt Lake City, and many others around the country with whom I have worked over the years. Alberto left Loomis, and I lost track of him; I miss that big smile. Mike Madden finally retired from the DoE and put the great lock debacle behind him for good.

After nearly four decades of Blizzards and Dilly Bars, my folks finally sold their beloved Dairy Queen. Mom likes having more time for knitting and her great-grandchildren, and Dad enjoys running his own political discussion forum (Buzzard's Roost, at delphi.com). The two of them love having more time for Cribbage, Rummikub, and each other.

I am in the process of semiretiring from the field myself. I spend a few hours each day running the Safecracker Support Forum and enjoy the challenge of providing tech support to safecrackers around the globe.

Valerie did the opposite of retire. She ramped up and is currently the Amicus Coordinator for the Washington State Association for Justice. She writes appellate briefs and argues before the Supreme Court.

It's debate and moot court all over again, but with bigger words and higher stakes. The WSAJ is lucky to have her, and so am I.

UPDATE to the Postscript. While I was finishing this book, my father's health declined rapidly, and he passed away. To the end, Dad remained stoical and never complained, not even when he was visibly uncomfortable. And he remained humble, preferring circumspection to conclusory claims about almost everything, including his own prospects.

For the past forty years, we exchanged books every Christmas, often on topics like the existence of God, the mind-body problem, and the possibility of life after death. In the late 1970s, he gifted me Raymond Moody's classic work, *Life after Life*, and I later got him Mortimer Adler's *Intellect: Mind over Matter*. More recently, we discussed at length the strengths and weaknesses in a Harvard neurosurgeon's account of his own Near Death Experience (NDE) — one that resulted in the surgeon's conversion from indifferent atheism to sincere theism.

Dad was fascinated by NDEs and loved exploring their implications. He was careful not to offend, but he didn't accept dogmatic answers from either side. He rejected the certitude of the organized religions that insist their particular interpretation of their particular holy book gives them capital-T truth, and rejected equally the certitude of the new atheists who assert with great confidence that there is no creator and absolutely nothing exists beyond the natural world. Dad suspected that people on both sides confuse what they so badly *want* to be true with the knowledge that it *is* true. Philosophers might one day call this the Is/Want problem.

He didn't go around beating people over the head with what he viewed as their dogmatic foibles. Nor did he deny that one side may well be right. The issue, as he saw it, was having or not having good reasons for thinking something is true.

Well, what *should* we make of near-death experiences and other phenomena that hint at something beyond? Dad didn't know, and neither do I. Maybe the reward is in the ride, and there is no after-party. Would that be so bad? This is, after all, a pretty great ride. On the other hand, perhaps the party doesn't end when the lights dim. Maybe it's just getting started.

As a son who would dearly love to play guitars with his father again, I'll confess to rooting for one side a little more than the other. Dad was a beacon, and we miss him dearly.

Acknowledgments

I OWE MANY DEBTS. TO MY PARENTS, FOR THE GIFTS OF LIFE AND CURIOSITY, for supporting my early and insatiable desire to pick locks, and for transporting me to and from Mr. Corey's shop during the teen years before I learned to drive. To Mr. Corey, for accurately pegging my mechanical aptitude along the dim end of the bell curve and forcing me to dig deep as a kid and develop workarounds. To author Steve Hamilton for asking me to proofread his award-winning book, *The Lock Artist*, and for encouraging me to put pen to paper. To my friend, Baksheesh Ghuman, for suggesting the day-by-day, blow-by-blow format of this book. To Kerri Altom, Jon Berlin, Pat Ekman, Martin Holloway, Joe Locke, Mike Madden, Lance Mayhew, Dave McOmie Sr., Justin McOmie, Ron Moore, Charles Santore, and Mike Swierzy for their many excellent suggestions. To Barrie Balter and Christine Pountney, for introducing me to useful concepts like psychic distance and metaphorical resonance, among others, and for forcing me to pick a side in the lively debate over the Oxford comma. To my son, Alex McOmie, for his editing chops, and for catching an embarrassing number of typos *after* I had gone through the manuscript with what I thought was a fine tooth comb. To my agent, Richard Curtis, for patiently prodding me to keep the focus on safes and vaults rather than veering off into philosophy, and for his guidance through the maze- and mine-filled world of publishing. To my editors, Rick Rinehart, Lynn Zelem, Jasti Bhavya, and the rest of the team at Rowman & Littlefield, for their judgment on matters big and small — from commas to chapter headings to indulging my peculiar preference for spelling out numbers that aren't fractions or decimels — and for taking a chance on an unknown author. And finally, to my wife, Valerie, for her constructive comments, for checking my memory on a few points of our long history, for putting up with my unintentional idiosyncrasies and their occasional (but quite intentional) displays, and most of all for sharing her life with me.

Notes

CHAPTER ONE: THE CALL FROM VEGAS

1. Diebold's footprint is actually much larger than just the United States. As I was finishing this book, Diebold shocked the bank service world by selling their cameras and alarms division to Securitas and acquiring Wincor-Nixdorf, and then rebranding their company as Diebold-Nixdorf. They now have a presence in more than a hundred countries. Even so, it will be a long time before the new name rolls easily off my tongue.
2. To my chagrin, my beloved Beemer case is now obsolete. I need to make this one last until I fully retire.

CHAPTER THREE: FROM AIRPORT TO AIRPORT

1. The best of these are compiled in Bob Rossberg's fascinating book, *The Game of Thieves*.
2. Zero Halliburton's Elite series briefcase can be seen in dozens of movies. I used one for a while but had the surface-mounted latches pop open on me. My scopes didn't fall out, but they could have. So I switched from the Elite to the Premier case, which has flush latches that cannot be lassoed by the strap of another bag in the overhead compartment.

CHAPTER FOUR: HITCH AND THE VAULT

1. This particular banking chain refers to the entire facility as a "cash vault." To avoid confusion, I am calling the facility a currency center. Among the many sections inside the facility is a room with two-foot-thick walls and a three-ton door. That room and its door are what I refer to as the "vault."

CHAPTER SEVEN: ALBERTO AND THE ATM

1. Season two, episode six: "Peekaboo."
2. By my count, Spooge called his partner "skank" twenty-four times before she nonchalantly got her revenge.
3. There is, however, a clever front-door defeat of the first generation Cencon via its keypad: www.wired.com/story/atm-lock-hack-electric-leaks/.

I'm a fan of electronic safe locks, primarily for their ease of use. But nearly all of them have vulnerabilities that can be exploited. I have two different interrogators that I use to avoid drilling whenever possible.

CHAPTER EIGHT: TEMPTATION

1. See http://articles.sun-sentinel.com/2003-03-05/news/0303050152_1_commercial-burglaries-federal-sentencing-guidelines-john-mark-collins.

Regrettably, it appears that Mike got out of prison and did it again. According to *The Atlantic*, the Hernando Country Sheriff's Office held a press conference on December 12, 2019, in which they accused Mike and his crew "of 23 jewelry-store burglaries across the state of Florida, with a total haul of $16 to $18 million." The Sheriff described Mike as the "mastermind" and "one of the top five safecrackers in the world." The Sheriff is half right. Mike may be a mastermind with nerves of steel, but skill-wise he is a journeyman, nowhere near elite status. See www.theatlantic.com/technology/archive/2019/12/rise-and-fall-all-star-heist-crew/602977/.

2. See www.upi.com/Archives/1993/08/07/Pink-Panther-jewel-thieves-begin-prison-sentence/1314744696000/.

CHAPTER NINE: INVERTED MOTHERLODES

1. Preternaturally positive though she is, my mother's sense of humor doesn't reach quite as far as mine, which is a good thing. Her limits helped circumscribe my own, though not without a couple of envelope-pushing incidents that caused her to worry for a while. Here are two brief examples.

Mom and Dad went to dinner with friends and left us kids home alone. Initially, it was great. We had pizza, Coca-Cola, a functioning color TV, and no prison guards. I also had my first set of lock picks and a few old locks to practice on. But after boob-tubing and munching out for a couple of hours, my sisters became bored. And after I picked all the locks multiple times, I too was restless.

Then I got an idea. An awful idea. I got a wonderful, awful idea!

I grabbed one of our old butcher knives, broke off most of the blade, plunged it through my t-shirt with the handle sticking out, and duct-taped it to my chest. I then poured ketchup on and around the knife's apparent point of entry while my sisters turned out all the lights in the entire house.

A few minutes later, headlights beamed through the windows as my parents' car motored down the driveway. Debbie, Deanne, and Denise hid in the bathroom, while I laid on my back in the hallway, around the corner from the front door. There was "blood" everywhere, and my half-lidded eyes were frozen in a death stare.

As the adults came through the front door, I heard my mother ask, with obvious concern in her voice, "Why are all the lights out?"

"Something's going on here," my father said sternly, and the four of them rounded the corner toward the kitchen and what was going to be a game of pinochle. They nearly tripped over my broken, bloody corpse, with the butcher knife sticking straight up out of my chest.

My mother's friend, who had been my sisters' second-grade teacher, let loose with a scream that could have raised the dead. She scared the hell out of me, and I shot up off the floor. That freaked them out, and they jumped backward, away from me, and she screamed again.

It was comical to watch the adults transition from shock to grief to relief to anger, and back to relief, all in mere seconds. I don't recall how long I was grounded, but it was worth it.

Another prank went awry a year later. Before buying the Dairy Queen, Mom was a meat wrapper at a supermarket about a mile from the house. The butcher was a gentle giant of a man named Turl, who stood six feet, seven inches tall and weighed close to three hundred pounds. I was visiting Mom at work one afternoon when Turl's wife called with bad news: their car had broken down, and she couldn't pick him up. Could he catch a ride home with us? Mom, of course, said yes. Quitting time was in forty minutes.

Our family car was an old Chrysler with detachable leopard print seat pads. A few weeks earlier, my Dad had slipped a half-inflated Whoopee Cushion between the pad and seat. I came out after working all day Saturday at Mr. Corey's shop, climbed in the passenger side, and sat on the surprise. It went off loudly and Dad cackled. A lover of bathroom humor, I laughed too, even though the prank was on me.

With visions of replicating this gag, I ran all the way home, borrowed the Whoopee Cushion from the bottom drawer of Dad's nightstand, and jogged back to the supermarket. I inserted the half-inflated star of the show under the passenger-side seat pad, and waited patiently in the backseat, pretending to read.

Soon after, Mom got in the driver's seat and closed her door as Turl opened his. He folded his enormous frame and sort of fell with style into the bucket seat. He hit the center of the bulls-eye and ripped a juicy one, complete with a little mouse squeak at the end. I was about to hoot and howl when Turl, face flushed in anguish, looked over at Mom. "Oh my god, Wilma," he stammered, "I'm so sorry," and quickly rolled down his window. *He thought it was real!*

Mom, ever the mitigator, replied instantly, "Oh, it's nothing I haven't heard before, Turl. No big deal." *She too thought he was guilty!* Only one person in the car knew the flatulence was fake. Not knowing what to do, I did nothing. I spoke not a word on the drive home.

That evening I told Dad the whole story. He laughed so hard and for so long that he had to remove his glasses and dab his eyes with tissue multiple times. When we later told Mom, she just stared at me, hands on hips, head tilted sideways. Her eyes were scrunched in the same uncomprehending expression I had seen a year earlier, with the butcher knife duct-taped to my chest. She just didn't see the humor. But she still loves me.

2. My oldest sister Julie died from illness as an infant, and my youngest sister Denise from an ocean accident as a teenager.

CHAPTER ELEVEN: BANK OF AMERICA'S UNTUTORED TENET
1. On page two of their "Vault Lock-Out Procedure" document, Bank of America states that "Coring the vault wall is the preferred method." On the same page, they explain why: if the vault door itself is drilled, BofA claims its "UL rating will be lost." This is demonstrably false, as the next two endnotes will make clear.
2. In 1992, I interviewed Lanny Gray, UL's Engineering Group Leader, who added to the confusion by making a series of remarks that were either unclear or contradictory. Over the next decade, I interviewed him several more times for my safecracker magazine and pressed hard for UL to clarify their public position on this issue. (These interviews were compiled into two articles that appeared in *The National Safeman* (which later became *The International Safecracker*). See Autumn, 1992, pp. 27–29; and Autumn, 1995, pp. 16–20.) UL eventually took the only position that made sense. The full text is in the next endnote.
3. In October of 2006, UL posted the following clarification on their website, where it remains today:

> *An authorized use of the UL Mark is the manufacturer's declaration that the product was originally manufactured in accordance with the applicable requirements when it was shipped from the factory. When a UL-Listed product is modified after it leaves the factory, UL has no way to determine if the product continues to comply with the requirements used to certify the product without investigating the modified product. UL can neither indicate that such modifications "void" the UL Mark, nor that the product continues to meet UL's safety requirements, unless the field modifications have been specifically investigated by UL.*

This was helpful. I followed up with more questions and later received this further clarification from UL's Regulatory Services Department:

> *Dave, strictly speaking, drilling does not "void" the UL Listing. The Listing Mark means the product met UL's requirements for Listing **when it left the manufacturing site**. However, once it has been modified (i.e., drilled and repaired), it may or may not still meet our requirements. We would have no way of knowing one way or another, as we were not party to the drilling and repairing process. (Emphasis in original)*

4. These are doors that were made before UL began testing vaults. Interestingly, the pre-UL doors are far more secure than any UL-listed vault door being manufactured today, and it's not even close.

CHAPTER TWELVE: A CHARTERED FLIGHT
1. The original design was from Herring-Hall-Marvin (HHM), and used in one version of their 5-Star Constellation vault door. Diebold bought HHM in the 1950s and adapted the design to their own door, the Guardian.

CHAPTER THIRTEEN: AT THE ESTATE: TWO TOUGH SAFES

1. In the trade, the term "tumbler" was replaced by "wheel" decades ago, and in technical writings I abide by the convention. Elsewhere I prefer to use the more mysterious throwback.

CHAPTER FOURTEEN: SAYING GOODBYE TO MR. COREY

1. Teddy also introduced me to Shintoism, the religion she was raised with and still practiced in her own quiet way. This was my first encounter with *real* religious diversity, far beyond the mere in-house disputes among Christians over peripheral issues that surround a fairly solid core of agreement.

Shintoism, with its lack of a holy text, denial of the Almighty, and strong emphasis on spooky sounding spirits and rituals, might as well have come from Mars. It was totally alien to me, and, for that reason, utterly captivating.

Teddy felt the same way about Christianity. She was fascinated but deeply skeptical. And the more I told her, the wackier it sounded. Salvation, atonement, and virgin birth were difficult enough, but the Trinity left Teddy shaking her head as if I had said two plus two equals five.

"How can a father be his own son?" Teddy asked, wrinkling her nose. "So sorry, David, but that doesn't make good sense."

Likewise, the more she told me about Shintoism, the wackier *it* sounded to *my* ears. "Why do you believe those things?" I finally dared to ask, wide-eyed and curious.

"Probably for the same reason you believe the things you believe, David," she said, as a matter of fact. "My parents taught me. It became a part of my life."

Light bulb moment for the young apprentice. It was blindingly obvious that if I had been born in Teddy's shoes and lived in her home with her parents in her culture, I would have been a Shintoist rather than the Christian I was at the time. And, of course, vice versa.

Wacky became relative, and I was impelled to reexamine those things I thought were true, to try and ground them in something more than lineage or turf. That's a tall order, and I've yet to fill it to my satisfaction.

So often I find myself pressing a diamond-like drill against the door that divides us from the ineffable, driven by a relentless yearning to penetrate part of the mystery, to glimpse a flicker of truth. So far, for me at least, the flickers have been faint, few, and far between. Wishing is easy. Knowing is hard.

CHAPTER SEVENTEEN: THE GREAT LOCK DEBACLE

1. As I was finishing this book, I was called out to the DEA field office in Yakima, Washington, to open a government safe. Take one guess which lock was on it. Yep, an X-07. I opted to drill it conventionally rather than glue it open, to avoid questions from the special agents who were onsite with me.

CHAPTER NINETEEN: DAD AND MORTY

1. Interestingly, my daughter Jennifer's college story parallels Adler's. Adler earned a PhD but was denied a bachelor's degree because he refused to take the mandatory swim class at Columbia University way back in the 1920s. Crazy but true. Similarly, Jenny is one of the few MDs in the United States without a high school diploma. She spent her junior year at a Catholic school in Argentina, and for some odd reason the American high school back home refused to accept the transcripts and wanted her to repeat the eleventh grade. We sent her straight to college instead. She graduated at twenty with a bachelor's degree in political science, interned at the state capitol, and quickly discovered that, while she loved abstract political philosophy, she didn't enjoy the sausage-making aspect of real-world politics.

She decided to become a pediatrician instead, which meant two more years of undergrad just to fulfill the science requirements for applying to med school. To top it all off, Jenny lived on four different continents for at least a year each before she turned thirty. I'm yet to leave continent number one. I hear the bravery gene sometimes skips a generation.

2. In high school, long before the VCR came along, I would hold an audio cassette recorder next to the TV speaker and record episodes of *Firing Line* to listen to in bed after the lights were out. In those days, not having the video was a partial deprivation, sort of like skipping an appetizer, because it was so much fun watching Buckley lean back in his chair, flick his tongue, and raise his eyebrows while readying a riposte. But the main course was always aural — listening to a gifted architect of language compose verbal works of art on the fly and deliver them in that evocative, hard-to-place accent.

In terms of polemical panache, Buckley's real rival wasn't Gore Vidal or Noam Chomsky, but Christopher Hitchens, especially in the 1980s and 1990s before Hitch began moving to the right on certain issues. I never had the pleasure of meeting either man, but like many fans of smart and witty discourse, I miss them both, may they rest in peace.

CHAPTER TWENTY-TWO: SKIP'S MAXIM

1. See http://old.post-gazette.com/localnews/20020606vault0606p4.asp.

CHAPTER TWENTY-FOUR: PLAN C: THE CLOCK IS TICKING

1. The thermal lance is a real tool, commercially available in most welding supply houses. I have owned several different lances over the years, and they are impressive to watch in action. The tip of the three-foot rod burns at nearly eight thousand degrees Fahrenheit and melts through steel and concrete like butter. Strangely, it struggles to penetrate wood!

2. The water bomb is also real, but no professional safecracker would ever use one due to the extensive damage caused by the powerful explosion. The idea is simple and ingenious: drill a hole into the top of a safe. Seal the cracks around the door jamb and then fill the safe with water. Finally, drop a blasting cap into the hole, seal it off, and detonate. The

force has to go somewhere, and that somewhere usually results in the door being blown off the front of the safe. Such an attack was used several consecutive summers on safes in the northeastern United States, which led some to believe the culprits were college students, perhaps graduate students from MIT. No one has yet confessed — or bragged.

Chapter Twenty-Nine: *BOMB! and Miscellany*

1. I'm not claiming that government handouts are always a bad idea. I'm saying that along with whatever good they do are real harms to *some* people who will give up and get by on the bare minimum rather than work their way into a better life. These are the invisible costs that never get counted, *because the victims themselves are likely unaware*. For that reason, they are easy for politicians and bureaucrats to ignore.

Chapter Thirty: *Thinking about Tomorrow*

1. For example, ABC already reported that the vault was drilled open — but I'm still at home waiting to make the trip to Paisley Park. Many media outlets copied ABC's story, but without crediting them as their source. Interestingly, ABC has offered no source either, other than to credit their local affiliate for having been first to report that the vault was opened. This is authentically fake news! See https://abcnews.go.com/Entertainment/prince-prescription-drugs-found-possession-home-law-enforcement/story?id=38719162, and https://abcnews.go.com/Entertainment/princes-vault-reportedly-drilled-open/story?id=38766251 and also https://consequenceofsound.net/2016/04/princes-vault-drilled-opened-and-theres-enough-music-to-release-a-new-album-every-year-for-the-next-century/.

Index